Springer Biographies

The books published in the Springer Biographies tell of the life and work of scholars, innovators, and pioneers in all fields of learning and throughout the ages. Prominent scientists and philosophers will feature, but so too will lesser known personalities whose significant contributions deserve greater recognition and whose remarkable life stories will stir and motivate readers. Authored by historians and other academic writers, the volumes describe and analyse the main achievements of their subjects in manner accessible to nonspecialists, interweaving these with salient aspects of the protagonists' personal lives. Autobiographies and memoirs also fall into the scope of the series.

Harsh K. Gupta

Harsh K. Gupta

Not A Day Different: My Life in Earth Science

Springer

Harsh K. Gupta
National Geophysical Research Institute
Hyderabad, India

ISSN 2365-0613 ISSN 2365-0621 (electronic)
Springer Biographies
ISBN 978-3-031-63560-1 ISBN 978-3-031-63561-8 (eBook)
https://doi.org/10.1007/978-3-031-63561-8

This Springer imprint is published by the registered company Springer Nature Switzerland AG
The registered company address is: Gewerbestrasse 11, 6330 Cham, Switzerland

To
my parents, my wife, daughters, and their families

and

all the wonderful people whom I met and got their love,
affection, inspiration and support that sustained me
in my life's journey

Foreword

It is my great pleasure to write the Foreword of *Not A Day Different* book authored by Prof. Harsh K. Gupta.

I have known Prof. Harsh K. Gupta for over 3 decades and have always been highly impressed by his outstanding accomplishments in Earth Sciences both at the international level and in his own country, India. Professor Harsh K. Gupta is internationally renowned for his pioneering work on earthquakes triggered by filling of man-made water reservoirs and on how they differ from tectonically induced earthquakes. His several other important research contributions include the seismic and geodynamic processes affecting the Tibetan Plateau, Himalayas, Bay of Bengal, and Arabian Sea, as well as the characterization of seismic rupture zones in stable continental regions, in particular in India. In addition to major contributions to geosciences, Professor Gupta played a remarkable leadership on the national (India) and international scene. In his own country, while serving at the highest level within the Indian government, he played a decisive role in developing major research centres (e.g., the Centre for Earth Sciences Studies in Trivandrum and the National Geophysical Research Institute in Hyderabad) and novel programmes in geosciences. The latter include projects dedicated to the groundwater resource for crop irrigation and food security, a topic of high societal relevance, as well as to hydrocarbons resources, for example, gas hydrates exploration in the Indian exclusive economic zone.

Professor Gupta's leadership is not limited to India. Over the years, he played major roles at the highest level in several renowned international organizations: among them the International Union of Geodesy and Geophysics: the Asia Oceania Geoscience Society where he served as the President, as well as the International Council of Scientific Unions (restructured as the International Science Council in 2017), the American Geophysical Union, and several others.

A decade ago, he edited the *Encyclopedia of Solid Earth Geophysics* published by Springer. The monumental success of this encyclopedia with over eighty thousand downloads by 2018 and the new developments in several Solid Earth Geophysics topics generated the demand to bring out a second edition. This demand was successfully accomplished in 2021 with the publication of the second edition of the Encyclopedia in two volumes, with 1950 printed pages, 260 articles written by over 300 authors globally.

The present book *Not A Day Different* summarizes the outstanding scientific achievements of an exceptional scientist with a remarkable personality. He has narrated how certain happenings, which appeared to be very undesirable, turned out to be a boon in disguise.

The book starts with a brief biography of early princely life of Harsh nested in a small town in north India, belonging to a very rich Zamindar (landlord) family. Abolition of Zamindar after India's independence and loss of fortune resulted in moving from Moradabad to Mussoorie, a hill resort in north India. The lifestyle changed totally and walking 12 kilometers every day from home to school and back, an hour of NCC training/boxing in the morning before the classes started, and another hour of swimming/playing field hockey after the classes were over, toughened Harsh to face difficult situation in the life. So, the abolition of Zamindari, which one would think as a curse, resulted in a boon.

One of the most outstanding examples of his leadership is setting up and populating Dakshin Gangotri, the first Indian wintering station in Antarctica in one Antarctic summer of 1983/84. This he did despite a near fatal helicopter crash of an MI 8 Helicopter soon after arriving in Antarctica. He believes, the helicopter crash, and subsequently his commitment to the then Prime Minister of India, Mrs. Indira Gandhi, that he would return from Antarctica only after setting up and populating the wintering station, motivated the expedition team immensely and succeeded in achieving this near impossible goal.

Early in life at the age of 25, in 1967 he discovered an unusually thick crust of 65–70 km below the Himalaya and Tibet Plateau region using earthquake generated surface waves. This first order discovery was confirmed in 1980s by detailed seismic sounding surveys. In 1967, he succeeded in establishing a seismological observatory at the National Geophysical Research Institute. Hyderabad, without an external help, similar to the World-Wide Standard Seismograph Network (WWSSN) set up by the United States Geological Survey, globally. He is known for generating criterion to discriminate artificial water reservoir triggered seismicity (RTS) from normal earthquakes, which are applied globally. He is also credited to have made successful medium term forecast of a magnitude 7.3 earthquake in the north-east India region.

During the 1990s, Harsh chaired the Steering Committee of the Global Seismic Hazard Program (G-SHAP), where some 500 scientists worked globally from 1992 to 1999 and produced the first global seismic hazard map. After the disastrous 26 December 2004 Sumatra earthquake of Mw 9.2 and the resultant tsunami, that claimed over 250 thousand human lives in the south and south-east Asia, Harsh spearheaded setting up of the Indian Tsunami Early Warning System for the Indian Ocean. This was completed in 2007. One of the highlights is placing ocean-bottom-pressure recorders in the vicinity tsunamigenic earthquake source zones to avoid false alarms. Harsh has been involved in several globally important initiatives such as setting up of the Asian Seismological Commission, Chairing the ICSU's Regional

Office for Asia and Pacific and preparing Science Plans for mitigating natural hazards in the Asia-Pacific region as well as addressing the Special Vulnerability of Island Nations to natural Hazards. Lately, Harsh has been busy in trying to develop earthquake resilient society. It is estimated that if one of the past great earthquakes in the Himalayan belt repeats today, several hundred thousand human lives could be lost.

To conclude, I would like to thank Prof. Gupta for having spent time and effort to write such an exciting book, offering us a narration of an extraordinary life and of scientific achievements that were not easy to achieve. The book will attract several generations of scientists involved in geosciences and certainly will inspire the youngest generation.

As a final word, let me recall the date of 17 December 2008, as on that day. Harsh Gupta received the prestigious Waldo E. Smith Medal of American Geophysical Union. I had the pleasure of nominating him for this award and was very happy for the success of this nomination. After the initial citation and Harsh's response to it, I had asked him a question, "*Harsh, if you had a choice, what would you like to change in your life?*", and his response was "*Not A Day Different*". I am so happy to note that Harsh has penned an autobiography with the same title.

A. Cazenave

February 2024

Anny Cazenave
Emeritus Scientist at LEGOS-CNES
Toulouse
Member of the French Academy
of Sciences
Toulouse, France

Acknowledgements

I am very lucky to have met wonderful, caring, and supporting individuals all my life. But for their time, love, training, and guidance, it would not have been possible for me to achieve whatever little I have achieved. The individuals who influenced and helped me, from the childhood include my first teacher, Mr. Ram Nivas; and later teachers Mr. H. N. Mehrotra, Rev. Brother Ryan, Prof. J. Singh, and Mr. B. P. Saha, Dr. B. P. Radhakrishna and Mr. V. Ramachandran, Dr. V. P. Dimri. Dr. Hari Narain deserves a special mention for carefully nurturing and supporting me scientifically. Interactions and working with Prof. Yasuo Sato, Prof. Mark Landisman, Prof. Anton Hales, Prof. Jim Combs, Prof. James Carter, Miss Frances Delany, Prof. Claude Froidevoux, Prof. S. Duda, Prof. Kiyoo Mogi, Prof. Tetsuo Santo, Prof. Anny Cazenave, Prof. Mioara Mandea, Prof. Rainer Kind, Prof. Bob Engdahl, Prof. Seiya Uyeda has been a pleasure. Several of my colleagues, too many to name, wherever I worked in India, USA, Japan, France, Germany, and elsewhere were always kind and helpful.

A special thanks to Anny Cazenave for the Foreword.

While writing this autobiography, my colleague R. K. Chadha has helped me a lot from the very beginning till the end. I am indeed grateful to him. At times, useful inputs were provided by Kusumita Arora.

The idea of doing this autobiography was originally discussed with Petra van Steenbergen of Springer. She introduced me to Margaret Deignan of Springer, who has been very helpful in developing the proposal that has finally produced this autobiography.

This work has been completed while availing Indian National Science Academy's Senior Scientist Fellowship at the CSIR-National Geophysical Research Institute, Hyderabad, India. I am grateful to the Director and staff of this institution for all the support. Uma Anuradha helped me with the preparation of the manuscript.

I have been fortunate to have wonderful parents, a very understanding wife, Manju and two lovely daughters Nidhi and Benu. They always stood by me in all my endeavors.

Harsh K. Gupta

Author's Note

This book is written to communicate with my fellow seismologists, researchers, and the young students who have chosen research as their career. Me not being a novelist or a storywriter, there may be shortcomings in my writing. However, having devoted my entire life in trying to understand the mysteries of Mother Earth through various geophysical tools, specifically earthquake generated seismic waves and genesis of earthquakes, a wealth of experience has been accumulated. Today, I can claim to have the practical and contemplative understanding that comes to a dedicated, long-standing goal-oriented practitioner of scientific research.

The thought to write my memoirs started crystallizing in my mind in 2022 while completing 80 years of a very eventful life that I had. I thought that it is time to tell my success story that may inspire the youth in my profession and others too, by sharing my experiences, contributions, dreams, and understanding of the subject and the opportunities it offers to serve the humanity. As a scientist, I had earlier used the art of communication to propagate my ideas and research on my understanding of earthquakes in several research papers in reputed journals and books, both at national and international levels but seldom presented my side of the story as to how I fulfilled my commitments to the nation while carrying out my research. This might have been done so, occasionally in a fragmented manner during some small talks in post seminar or symposia dinners or some meetings or in social gatherings but never in a comprehensive and organized manner.

Reading through this book, it may perhaps become obvious to the readers that not only the contributions made towards setting up of national assets have been mentioned in the book, but future direction of work is also outlined, specifically on earthquake related matters, also pointed out the future directions of seismology research in the country. I am sure that still there is a long way to go where new chapters of accomplishments will be added by the younger generation in developing a strong scientific aptitude to take the world forward.

It may be noted that this book is a communication with myself as it is with the readers, and I am aware of my limitations and weaknesses. There may be several incidents and occasions that may have been overlooked or have not made any attempt to gloss over them. Hopefully, the readers will find an author who is honest with them and with himself.

In the last, it may be mentioned that all my life I have been writing research papers and theme-based science books that, once they are written and published, belong to the readers as to the author. A list of all my research publications has been placed at the end that will serve as an archive for reference at one place and can be used by anybody pursuing seismology research.

Hopefully, this book shall succeed in communicating something meaningful to the readers and they shall find it interesting.

Harsh K. Gupta
Hyderabad, India

Prologue

Not A Day Different echoed my voice in the auditorium of the Moscone International Convention Centre and lost in the thunderous applause. The occasion was the award ceremony of the prestigious Waldo Smith Medal conferred on me by the American Geophysical Union in 2008 in San Francisco, USA, and a question was put to me, by my nominator, Anny Cazenave, "*Harsh, if you had a choice, what would you like to change in your life?*"

This award was given to me 25 years after its inception in 1982. This award is given in recognition of extraordinary service to Earth and Space science by a scientist who has strengthened and helped advance our scientific disciplines, as well as played unique leadership roles in scientific associations, education, legislation, research, management, philanthropy, or the public understanding of science. In my acceptance speech, I went down the memory lane of a well-lived privileged life since my birth in the sleepy township of Moradabad, close to the Himalayan foothills in United Province of British India in a very well-to-do Zamindar family.

On 28 June 2022, I completed my first 80 years of life and while taking a brief rest, I reflected upon 80 years of my blissful journey spent in the company of some wonderful people who enriched my life with their kindness, trust, and faith in me. It was very humbling. I have been fortunate to occupy some very vital positions that are not in the normal reach of a scientist. I did not possess the advantage bestowed by factors such as my birth or position of my family in the society, nor did I seek political backing for reaching these positions. But, luck plays its role in everyone's life, and I had perhaps a larger share of it than others.

Today, when India commemorates 75 years of independence and the glorious history of her people and contributions to science, arts, and culture, I see my small footprints embedded somewhere in the development of seismology research in India, which also has a global impact. This gave me an incentive to write this memoir for the benefit of the generations to come. It gives me immense personal satisfaction that in my own special way, a contribution has been made to my motherland by putting

India on the world's map because of a few scientific contributions with which I had been involved.

In a nutshell, it can be said that my life has been a very active one and it continues to be so, even today. Though, a very short one in comparison with the history of our country, my journey from 1942 till now is quite considerable in an individual's life. The seismological research in independent India has subsumed my professional journey giving me an opportunity to both observe and in my own humble way contribute to many momentous developments along the way. I led an eventful life brimming with activities and full of vicissitudes, however, highly satisfying in totality. It has helped me evolve into the person that I am today.

Receiving Waldo E. Smith Medal for 2008 from the American Geophysical Union President Timothy Grove (right).

2008 Waldo E. Smith Award function in San Francisco. From right: Alik Ismail Zadeh, Anny Cazenave, Harsh K. Gupta, Manju Gupta, and a guest.

Contents

About the Author

Harsh K. Gupta Senior Scientist, Indian National Science Academy; President, Geological Society of India; Member, Atomic Energy Regulatory Board, India.

Dr. Gupta has published over 200 scientific papers in internationally reputed journals, has authored 5 books, published by Elsevier and Springer, and edited 21 volumes. One of his papers was adopted as a chapter in the Open University (U.K.) text- book on Geophysics. His first book, "Dams and Earthquakes", published in 1976 was translated into Russian and Chinese languages. Prof Gupta compiled and edited "Encyclopedia of Solid Earth Geophysics". This monumental 1500+ pages two volume treatise, published by Springer is globally a landmark (website: www.springer.com). The second edition of this Encyclopedia (1950 pages, two volumes) has been published by Springer in 2021.

Important positions held by Dr. Gupta include:

Member of the National Disaster Management Authority of India (Status: Minister of State, Government of India, 2011–2014); 2) Secretary to Government of India, Department of Ocean Development (2001–2005); Director, National Geophysical Research Institute, Hyderabad (1992–2001); Advisor, Department of Science and Technology, Government of India (1990–1992); Vice Chancellor, Cochin University of Science and Technology (1987–1990); Director, Centre of Earth Science Studies, Trivandrum (1982–1987); Adjunct Professor, University of Texas at Dallas (1978–2001); Research Scientist, University of Texas at Dallas (1972–1977).

Major Scientific Contributions:

Prof. Gupta's work is globally recognized for 1) Discovering an extremely thick 65–70 km crust below Himalaya and Tibet Plateau region, 2) Developing criteria to discriminate artificial water reservoir triggered earthquakes from normal earthquakes, which are globally applied and finding safe sites for construction of reservoirs, 3) Making a medium term forecast of an M~8 earthquake in north-east India region in 1986 which came true on August 6,1988, 4) Chairing the Steering Committee of the Global Seismic Hazard Assessment Program (G-SHAP) where some 500 scientists worked from 1992 to 1999 and produced the Global Seismic Hazard map, 5) Pioneering the Gas Hydrate program and delimiting the zones for stability of gas hydrates within the exclusive economy zone of India, 6) Taking up detailed studies of genesis of triggered earthquakes in the Koyna region and making successful short term earthquake forecasts.

Setting up Dakshin Gangotri:

Dr. Gupta was the Leader of the 3rd Indian Scientific Expedition to Antarctica. The Expedition was charged with the responsibility of setting up and populating the first Indian wintering base in Antarctica in one Antarctic Summer of 1983–84. No country had done so till then. The 3rd Expedition succeeded in setting up of Dakshin Gangotri and populating it despite having a near fatal MI 8 Helicopter accident on the third day of their arrival at Antarctica. The station had a plinth of 620 sq. m and all the facilities and full-fledged laboratories for scientific work. This created a world record that still holds.

Indian Ocean Tsunami Warning System:

After the occurrence of the devastating 26 December 2004 Mw 9.2 Sumatra earthquake and the resultant tsunami that claimed over 250,000 human lives in south and southeast Asia, Prof. Gupta spear headed setting up of the Indian Ocean tsunami warning system, which was completed in just 30 months' time. This is now assessed to be among the best tsunami warning systems globally.

IUGG/ICSU/AGU/AOGS Etc.

Long time involvement with IUGG (President 2011–2015, Vice President 2007–2011, Bureau Member for two terms, 1999–2003 and 2003–2007); IASPEI (Vice President, 1995–1999, Executive Committee Member 1991–1995); Chair, Steering Committee of Global Seismic Hazard Program (GSHAP), 1992–1999; ICSU (Member CSPR, two terms, 2005–2008 and 2008–2011); Chair, Hazards Group, ICSU Regional Office for Asia and Pacific; ILP (Bureau Member, 1986–1989; and was made a Bureau Member of ILP for life in 1996); Founder President of the Asian Seismological Commission (1996–2000); IUGS (Councilor, 2000–2004); Vice President and President AOGS (2010–2013); Member Scientific and Technical Committee of ISDR (2009–15); Member AGU Committee on Public Affairs, The New AGU and a member of AGU Committee on International Participation, Member Science Committee of Deep Time Digital Earth (2019–continuing), etc.

Awards: Many, including:

1983	Shanti Swarup Bhatnagar Prize for Science & Technology in Earth Sciences
1985	USSR Academy of Sciences "100 Years of International Geophysics" Memorial Medal
1991	National Mineral Award, Government of India
1995	CSIR Technology Prize for Business Development and Technology Marketing
1998	Second Outstanding performance Award by the 11th International Kharazmi Festival of Iranian Research Organization for Science and Technology, Teheran, Iran

1999	Federation of Indian Chambers of Commerce and Industry, New Delhi, Award for Physical Sciences,1998–99
2003	Jawaharlal Nehru Visiting Fellowship 2003, Indian National Science Academy
2003	The 4th Prof. C. Karunakaran Endowment Lecture
2003	Bruun Memorial Lecture, 22nd session of the Assembly of International Oceanographic Commission, Paris
2003	National Mineral Award for Excellence-2002
2006	Padma Shri 2006
2008	Prof. Y. Nayudamma Memorial Gold Medal Award for 2008 from APAS
2008*	Waldo E. Smith Medal Award for 2008 from AGU
2016	Axford Gold Medal of the Asia Oceania Geosciences Society (2016)
2019	IUGG Honorary Fellowship
2021	Prof. W.D. West "Lifetime Achievement Award"
2022	Elected Foreign Member of the Russian Academy of Sciences
2022	Elected Fellow of the International Science Council
2022	National Lifetime Excellence Award by the Ministry of Earth Sciences, India.
2023**	International Association of Seismology and Physics of Earth's Interior (IASPEI) Medal.

*Prof Gupta is the first so far and the only scientist from the developing world to be awarded Waldo E. Smith Medal by the American Geophysical Union.

**Prof Gupta is the first one to receive this medal outside USA, Europe, and Australia.

Chapter 1
Early Life and Introduction to Earthquakes

1.1 Brief Biography

Quit India Movement 8 August 1942—I was hardly 42 days old when the final push for India's independence came when Mahatma Gandhi gave an ultimatum to British Raj to Quit India against the backdrop of World War II and launched a "do or die" agitation to attain *poorna swaraj (total independence).* Those were testing times for every Indian and the atmosphere in the country was charged with great nationalist fervor. My grandfather and my parents saw to it that my early childhood was insulated from the fury of agitations and unrest in the country due to communal disturbances that became the order of the day. Growing up in such an environment where every Indian was yearning to live in a free country, perhaps, the first seeds of a fiercely independent and a strong self-reliant individual were sown in my young mind with a pride that reflected the true spirit of those times.

I was brought up in great comforts and a secure environment of our sprawling bungalow *Hari Villa* that boasted of a beautiful lawn and a garden with a 5 ft deep lily pond in the center in which colorful ornamental fish swam elegantly. The district of Moradabad, not very far from the foothills of the mighty Himalaya got its name after Murad, the son of Mughal Emperor Shah Jahan. The city is world famous for its brass handicraft's export to Europe and North America and got its nickname "Peetal Nagri" (Brass Town). My grandfather, Sahu Ram Kishore was a Zamindar in Thakurdwara tehsil, one hour's drive from Moradabad where he moved in 1935. My father Sahu Vishnu Kumar and my mother Sushila Devi provided all attention to our family of four siblings, an elder brother and a sister and a younger brother. We had an additional member Shera, a dog of Alsatian breed brought by my father when I was born. In fact, tryst with accidents had been my birthmark that emboldened me to face challenges in life without a sense of fear. It all started when I was a few months old, I slipped out of my sister's lap while she was watching the colorful fish and fell in the 5 ft deep lily pond at our home *Hari Villa.* An ever-alert Shera, jumped into

H. K. Gupta, *Harsh K. Gupta*, Springer Biographies,
https://doi.org/10.1007/978-3-031-63561-8_1

the pond immediately and rescued me in an unconscious state as substantial water had entered my lungs (Photos 1.1 and 1.2).

Every child inherits certain characteristics that reflect the socio-economic and emotional environment into which he or she is born. Early memories of India attaining independence were both joyful and grim. I vividly remember Moradabad being lit up with earthen lamps on 15 August 1947 and there were celebrations all over. I also recollect the sufferings of people after the mass exodus of Hindus, mostly Sikhs

Photo 1.1 Early childhood photograph, from left: my father Vishnu Kumar, brother Anil, sister Saroj, grandfather Ram Kishore, with me (the baby)

Photo 1.2 Hari Villa, Moradabad, India, where I was born and grew as a child

and Punjabis, from the newly created Pakistan. A few thousands of them finding their destination to our city. Some of them, especially hunger-stricken widows and children who took shelter in a refugee camp near our *Villa* left an indelible mark on my psyche as a child. Unable to see their plight my mother would get extra food cooked every day and serve them. About 20 of them would come and fed by my mother. They request to provide them with some work to do in lieu of such a kind gesture of my mother as they refused free food. They never begged and accepted their destiny in the larger interest of the country. The strength of their character greatly impacted me, and I very often think about them as the real unsung heroes who never gave up even during extreme adverse times. It was not surprising that these proud women soon found an honest occupation, their number dwindled, and within six months, none would come for food, but once in a way visit my mother to express their gratitude.

Looking back, I realize that a *"never to give up"* attitude from these brave people at a very early age in life was developed. My mother and father were very kind and honest people. My elder brother Anil was very disciplined who always excelled in his personal and professional life. Their conduct and behavior deeply influenced my mental development. Armed with such traits I entered my youth as a very sensitive youngster, at the same time being extremely conscious of achieving the set goals and not hesitating in taking chances for achieving them.

1.1.1 Formative Years

Early education during pre-independence era was at home, where Mr. Ram Niwas, a railway employee and an excellent teacher, used to come and teach everything from languages to science, social studies, and math, except Hindi for which there was another teacher. I still remember writing Hindi on a Takhti, a 1×2 ft wooden plank with a bamboo pen and earthen ink pot that later used to be wiped clean using mud paste for the next day, a practice followed in those days in regular schools, too. After lunch, I use to run to our backyard in *Hari Villa* that hosted several families appointed to take care of the maintenance of our home. I remember the stories told by Munna Lal, one of our helps, who made us believe that his uncle has a ghost friend who is not visible to anyone except him and helps him whenever he needs him. Time flew fast and as I turned eight, my time at fun school at home was up and I was to head to a regular school to explore the world, outside the comfort zone of our *Hari Villa.*

In 1949, two years after the independence of India my teacher at home Mr. Ram Niwas advised my father that it was time I get into formal education in a school. I was taken to the prestigious Montessori School in Moradabad, a Primary School that imparted education up to Class V and was headed by an eminent teacher trained in Ceylon (Sri Lanka), Mr. Sarveshwar Saran Serve. After a little conversation with me he advised Mr. Ram Niwas that I was more suitable for Class VII, and he may consider putting me in a Middle School. In those days, there was no age limits for admission to schools. Mr. Ram Niwas narrated what happened at the Montessori School to my father. But as a doting father, he felt that I was too young to be in a higher class. So, my formal schooling started at a Montessori school when I was admitted in Class V. My days at the school were full of pleasant experiences as we were only 12 students in the class V, and a total of only 70 children in the entire school. There was so much of personal attention and teachers were very affectionate and caring. One of my cousins, Yugeshwar Saran Kothiwal was my classmate, with whom I have remained in contact till today. During my first formal schooling, a national tragedy struck the country. The father of nation Mahatma Gandhi was assassinated on 30 January 1950. I still remember when my entire family was glued to our Westinghouse Radio at home listening to the commentary in progress of the procession carrying the body of the Mahatma Gandhi to Raj Ghat, his final resting place in Delhi. There was so much uncertainty and chaos in the country that the very existence of India as a united entity was in doubt. Thanks to the wisdom of leadership at that time and maturity of the people of India that everything remained under control.

After my one-year stay at the Montessori school I was admitted to Class VI in the school section of Hindu College in Moradabad in 1950, where my elder brother Anil was also studying. Being a topper in school he was expected to repeat the performance in Class X public examination. The school had a large strength with 5 sections in each class of 40 students. The school had a well-organized infrastructure with each classroom equipped with comfortable desks and ceiling fans. There was a big playground for PT exercise and games, and a daily routine was prescribed for the overall development of students. The stress was more on imparting quality education rather than spoon-feeding that reduced the burden of carrying huge loads of books every day to the school. I had my first competition with myself as the burden of expectation to perform at the highest level was on me because of my brother's performance as a topper of the school. I was also expected to achieve the same level of success.

In this process, my brother turned out to be my mentor at home and became a good friend as we spent more time together. It was a pleasure going to school with him, which was about 10 minutes walk from home. As expected, Anil topped the school in the U. P. Board High School examination. And to the glory of our family and Hindu College, he also topped the entire ~ 3 hundred thousand students who passed the U. P. Board High School examination in 1951 and created a new record. Anil moved to Meerut College in Meerut for Intermediate education. While studying in Meerut, he came home during Durga Puja vacations and summer holidays and we, together indulged in some adventures like fishing with the help of Munna Lal, our help at home. I recollect my excitement of catching a fish, which we gave it to Munna Lal as *guru dakshina*, as we were vegetarians.

I was in Class VIII when the Government of India enacted "Uttar Pradesh Zamindari Abolition and Land Reforms Act 1950", on 1 July 1952 that ended the Zamindari era in Uttar Pradesh. My father tried his hand in a few businesses, initially but without much success and we had to undergo a brief period of uncertainty and struggle. In my own small way, I remember to have tried my hand at a business venture along with my brother Anil who came home during summer holidays. We planned a short-term ice cream business in the Moradabad Ice Factory in which my father had partnership.

There was a huge tank where water was cooled below 0 °C and there were 1 × 2 ft partitions where huge cans of 5 ft length filled with water were lowered and the water froze in sub-zero degree temperature. These cans were lifted by a crane and ice slab were recovered. We first understood the process and got a couple of 1 × 2 ft slots and used them for making ice-creams on a stick. The entire process took approximately 4 hours and after decanting, we employed two young boys to sell product in icebox carts commonly used by hawkers. It was an amateurish attempt, but nevertheless, it made me realize another facet of my persona at an early stage of life that later metamorphosed into great self-confidence.

Later, several responsibilities fell on my father, but it did not deter him from taking few very bold decisions that included selling our *Hari Villa* for Rs. 70,000/ - that was a princely amount in 1954 and move to our ancestral home *Ram Niwas* in Thakurdwara. The decision to sell the house was very emotional and against the wishes of almost the entire family but he stood his ground and backed his decision for the better future of the family. Today, when I look back to my childhood days, I realize how right my father was, and this singular action of his helped us tide over a brief period of struggle and uncertainty and bounce back in business with happy ending. I learnt a great lesson to be decisive and bold when faced with a crisis rather than run away from owning responsibility. *Fortis Fortuns Adiuat*, my teacher told me how Fortune favors the brave, while narrating the stories of the bravery of Greek, Roman and Indian legends.

After my parents shifted to Thakurdwara I stayed back in Moradabad to complete my Class X examination which I passed in flying colors, and it was time for me to move on to the next exciting journey of a college life.

1.1.2 Alma Matter

One's memories of early life as a child at home and in school are often nostalgic, invested with illusory romance. But viewing things objectively, I witnessed very early in life, a sense of great elation when India got independence from British rule and saw the darker side of partition of India and the miseries of the people. But in independent India opportunities beckoned with open arms to people who were educated, capable, intelligent, and willing to contribute to nation building. I cannot recollect a single reason for being dissatisfied with the conditions of my childhood or later during my college days as opportunities were waiting to be grabbed by well educated people who were ready to serve the country and take part in shaping India's destiny. In such a scenario my first steps of adulthood were taken in the St. Georges College, Mussoorie, situated in the Siwalik Mountains in the foothills of Himalaya.

From, St. Georges College, Mussoorie in the picturesque landscape of the Himalayan foothills to Indian School of Mines (ISM), Dhanbad in present day Jharkhand was my first opportunity and a big step in life. It was my first exposure to the outside world that gave me a sense of appreciation of the varied Indian landmass and cultural diversity of the country. We were staying at the Himalaya Club Hotel owned by our relative Dr. Ram Prakash Baba ji and was at a 50-min walking distance from my college, St. Georges. I joined NCC and boxing classes and would play hockey and swim on alternate days, after the classes. Playing Table Tennis and Badminton were past time on Saturday and Sunday evenings. My first brush with injury was when I broke my ankle during a practice session of Table Tennis. In early December, St. Georges used to organize a half-marathon (21 km) where some 150 students would run. Everyone, who completed, got a ranking. If 100 participants completed, the first will get 100 marks, 2nd would get 99 marks and so on. There was a prize for the class whose students added up to most marks. The Marathon started from St. Georges, then through King Kraig to Library to Charleville and then back. It was about 21 km. For the sake of my class, I ran and finished somewhere in 50's. This was just 2 months after my recovering from the ankle fracture. Staying at Himalaya Club I had another 6 km to walk before starting the Marathon from St. Georges and then walk back 6 km after finishing the marathon. The sports activities and my everyday 12 km walk from home to college and back helped me develop a good stamina (Photo 1.3).

One incident that would have taken my life was our misadventure while in Mussoorie. We planned a trekking expedition to Chakrata in Siwalik range of Himalaya during summer vacations when Anil came home. We started our 55 km walk in which my cousin Madan Mohan also joined. After covering half the distance, we reached a point where we must cross Yamuna River that flows majestically against a beautiful background of Himalaya, using the Yamuna Bridge. I told Anil that I will swim across the river and meet them on the other side while they cross the bridge. As

Photo 1.3 Me delivering chief guests address at the St. George's sports day in 2012. St. Georges College, Mussoorie, India is in the background

I entered the river and swam a little distance, I felt that the current was too strong and made me move along the flow. I tried very hard to swim cutting across the current and tried to reach the other side of the riverbank. This exhausted me and I somehow caught hold of a tree's branch bending into the river and almost fainted. Anil and Madan, waiting on the other side, rescued me before I lost my senses. It was a great lesson for me that day, to know that swimming in a pool is different than swimming in a flowing river, especially when you are swimming against a strong current. Though, a traumatic experience it taught me not to be shaken when experiencing difficult situations. This experience was greatly gratifying.

After completing my 12th Standard in Mussoorie, I moved to Allahabad University for my B.Sc. Degree. I was fortunate to be under the tutelage of several stalwarts like Profs. Gorakh Prasad, Pyare Lal Shrivastava, Deo Dhar and Rajendra Singh. There were 5 sections in B. Sc Part I, and all these stalwarts came to teach us without missing a class despite their heavy workload. Their knowledge, dedication and discipline made a lasting impression on me. My brother Anil who had graduated from IIT, Kharagpur and was working for ONGC told me about the prospects and opportunities associated in the field of Earth sciences, especially for Geophysicists and Geologists. Since, I was good with math and physics he advised me to take up Geophysics course offered at the prestigious Indian School of Mines (ISM), Dhanbad. The British Indian Government had established the Indian School of Mines, Dhanbad on the lines of the Royal School of Mines, London in 1926. In 1957 two more departments of Applied Geophysics and Petroleum Technology were added in ISM. Over the years, it had gained the fame of being amongst the top mining schools in the world.

I took my brother's advice seriously and wrote the All-India entrance examination of ISM at a Centre in Lucknow. Being an outgoing, adventurist and a curious person by nature I saw a great opportunity to unravel Earth mysteries that always fascinated me. It was delightful to have passed the entrance examination and get admission to a 4-year course in Applied Geophysics at ISM in 1959. The train journey in early July by Doon Express from Dehradun to Dhanbad is etched in my mind. With a dream in my eyes and great confidence and pride that I was going to be a part of an internationally reputed Indian School of Mines, elated me. The excitement made me feel that the train was moving very slowly. There was an intense hurry in my mind to reach my destination that was to decide my future, as a Geophysicist.

Prof D. N. Prasad, an alumnus of ISM was the Director in 1959. He was a reputed mining engineer with excellent connections to the industry. He would teach for at least 2 h every day and was the most sought-after teacher. Above all, he was a very kind man always coming to the rescue of needy students who very often exploited him by fabricating stories. There are several anecdotes where they were caught but he never took any serious action against any of them as he felt that these were the signs of a normal growing up process. Such a breed of teachers we miss today, and they ever remained in the memories of the students who were taught by them.

Before 1957, only Mining and Applied Geology were the two branches that one could graduate from at ISM. In 1957, Petroleum Technology and Applied Geophysics were added. These were 4-year programs with 60 students for Mining and 20 students each for Applied Geology, Applied Geophysics and Petroleum Technology with a total of 120 students in the first year. The total number of students in the entire ISM was less than 500! My admission number was 1473 meaning that I was the 1473rd student admitted to ISM since its inception in 1926, a remarkable achievement. For the students of all branches, the 1st year course was common. I was the only one with Applied Geophysics as my first choice in Section A of first year as the most preferred branches were Mining Engineering, Petroleum Technology and Applied Geology.

There are a few anecdotes of ragging at the hands of seniors. One of the most hilarious one was an ice-breaking event with seniors who became my best friends. One night after dinner, four seniors walked into my room and enquired from which place I had come. Learning that I am from Mussoorie, Dehradun they told me to climb on the cement almirah in the room. After following their command one of them told me to imagine that I have reached a mountain top in the hills in Mussoorie, and it is very cold there. There was no choice except to follow their command. Then they put a blanket on me and enquired if I still felt cold. Sensing the plot, I went ahead with their script and replied in affirmative. Then they put another blanket and the process continued till I was sweating in the sweltering heat of Dhanbad Coal belt. Seeing my response with no protest, whatsoever, and with a smile on my face they were all impressed by my sportsman spirit and hugged me and became my best buddies. Even for me it was a great revelation that I can be so cool under the most serious of provocation. The most wonderful aspect of being an ISM alumnus that made me proud even today is: whenever we ate a snack or had a cup of coffee with a senior; the senior always paid the bill, a wonderful tradition that continues even today.

The first lesson in management was learnt when I was made the Manager of our Mess in the hostel. In those days the monthly bill used to be around Rs. 60/- that included breakfast, lunch, dinner, and an evening snack. My two roommates K. L. Gupta and M. S. Chabbra assisted me. We used to plan the menu, took great pains to buy best possible grocery by ourselves for the whole month from the nearby Jharia Township to maintain quality. As a result, the monthly bill came down to Rs.55/month and everybody appreciated our efforts to provide quality food at a lower monthly expenditure.

It was indeed fortunate for me to have been taught by Prof Jagdeo Singh, Head of the Department of Applied Geophysics who was a brilliant teacher, mentor, and an individual with excellent human qualities. He would accompany us to fieldwork, typically of 15 days' duration and then would insist our completing the assigned tasks of making measurements every day. These field trips were full of fun, and we gained a lot of experience of real conditions. Every time Prof. Singh used to pay for our snacks and tea whenever we stopped at a place during fieldwork. Later, Prof. Deepak Chowdhury joined our department. He was the alumni of IIT Kharagpur and had obtained his Ph.D. from USA. He was an excellent teacher and joined hands with Prof. Singh to shape our future (Photos 1.4, 1.5 and 1.6).

Entering the Bhowrah Colliery was my first experience to look directly into the womb of Mother Earth, as a part of a compulsory training in the mines for two months after first year of the 4-year program, as a part of the curriculum. The Bhowrah Coal

Photo 1.4 Prof. Jagdeo Singh (Indian School of mines)

Photo 1.5 In 3rd year, a few of us won several trophies (me: 4th from left). Display of our prizes with Indian School of Mines main building in the background

Photo 1.6 Receiving M. Sc. applied geophysics degree from Prof D. N. Prasad, Director, Indian School of Mines

Mine is a very well established mine with a reserve of ~ 100 million tons of coal. This was a thoroughly enjoyable experience. My excitement kept increasing as I went deeper and deeper following the coal seam beds within the rocks. Several faults were visible and it was so interesting to see how they displace the strata and the coal seams. It was a great experience to see how coal is mined from greater depths of Earth and their pitfalls. Following a month long training on various aspects of coal mining, the second month was spent studying mica mining at the Koderma Mica mine. It was different than that of the coalmines at Bhowrah as this mine was primitive. There were no shafts to enter, and one must use an incline. The mining was not too deep, just about 50 ft from the surface. Entering the mine holding a candlestick was totally a different experience.

A chance meeting with the person who later made a great impact on my professional life is worth mentioning. Dr. Hari Narain, the then Director of Oil and Natural Gas Commission's Research Center at Dehradun was on visit to ISM on the invitation of Prof. Jagdeo Singh. Dr. Narain was a big name in Geophysics and had come to see the facilities and interact with the Geophysics students. At that time ONGC was a major recruiter of Geophysics graduates, however, there were not enough vacancies that can take care of all the students. As the spokesperson of the Geophysics students' group, I welcomed him over tea in the hall of the Old Hostel which he obliged smilingly and shook hands with all of us. That was rather unusual for us as our experience with people of celebrity status was just a wave of hand or a namaskar. Sensing an opportunity, I spoke for a few minutes then stressed on him that most of the students are trying to go to USA for advanced degree after the M.Sc. as there are not sufficient opportunities in the country. He listened patiently to all of us and then spoke a bit about ONGC and the challenges and opportunities in exploration of hydrocarbons in the country and the active role ISM can play. He told us that while he was not against students going abroad for obtaining higher degrees and experience, he would rather appreciate if they came back and serve the nation. The meeting lasted for about an hour, and we were happy with his concern and assurance that he would do his very best to see that employment situation for Geophysics graduates improves in the country. This meeting changed our outlook in our lives, and we all contemplated how best we could serve the nation in whatever profession we enter after finishing our studies at ISM.

Two incidents happened during the final year in 1962, one that almost took my life and the second was my selection for a position in seismology at the India Meteorological Department (IMD), New Delhi. As we were nearing our final days at ISM, we planned an excursion to the nearby Maithon Dam. We took a paddleboat to go around the reservoir where swimming was strictly prohibited. As we were enjoying the boat ride I was tempted to swim and told my friend Ashok Rudra who was with me on the boat to be around and keep a watch on me. After swimming for half-an-hour I started swimming back towards my boat when a high speed motorboat passed by between me and the paddle boat. The waves created by the motorboat pushed me farther away by almost 100 m. By this time, I was quite tired and realized that I was in no shape to swim back to the paddle boat. Good sense prevailed and I started floating on my back to conserve energy. Ashok, from a far distance was observing as to what happened to me and started paddling to reach me. By the time he reached me, he himself was very tired. Somehow, he managed to pull me into the paddle boat. This incident shook both of us but also taught several lessons of uncertainties of events that can happen without warning, and one must be prepared to face them.

While in the final year of M.Sc. Applied Geophysics program, I had applied for a position in the India Meteorological Department (IMD). An interview for this position was conducted by the Union Public Service Commission (UPSC) office in New Delhi. The results were declared towards the end of November 1962, and I was asked to join at the Central Seismological Observatory (CSO) at Shillong within a month's time. Prof. D. N. Prasad helped me in getting permission from IMD to join after my final examination in 1963. But, before joining IMD, I wanted to spend a few days at our newly constructed house '*Jeevan Madhu*' in Moradabad that reflected my father's resolve to bounce back to prosperity after going through several ups and downs in the rigmarole of life.

I must confess that apart from studies and sports, the bonds of friendship developed at ISM during my stay and the experience of a life of living together, travelling in rickshaw the only mode of transportation in Dhanbad and being away from home taught me several lessons that gave me confidence to face the world as a more mature adult. The journey as a seismologist to unravel and understand the mysteries of Earth was about to begin.

Abolition of Zamindari system post Indian independence, which appeared to be devastating for our family, when it occurred, turned out to be a boon for me, shifting to Mussoorie, studying at St. Georges, walking 12 km every day, in addition to sports and NCC, drowning/

near drowning 3 times, studies at the School of Mines, prepared me for the journey ahead full of adventure and capacity to survive in very demanding situations, that helped me all my life in achieving near impossible goals.

1.2 Initiation into Seismology

After spending a few days in our newly constructed home '*Jeevan Madhu*' in Moradabad it was time to step into my first employment in the picturesque landscape of Shillong in Assam. With a dream in my eyes and the excitement of entering an unknown world I boarded the train at Moradabad which took me to Guwahati. It was followed by a bus journey of 5 h from Guwahati to Shillong. The bus journey was very pleasant through the winding roads in the mountainous terrain of Assam to reach my destination at Central Seismological Observatory (CSO) of India Meteorological Department (IMD) located in a campus called "Meath Home". Mr. B. P. Saha was the Head of CSO who made me comfortable and introduced to other staff members. Later, Mr. Uday Shankar De, a geophysicist from Benares Hindu University who joined just one month before me gave a tour of our campus. I was fascinated by the lush green campus with rich fauna and flora overlooking a canvas of expansive mountain chains and valleys. Most of the staff spoke Bengali with few others Nepali and local Khasi languages. But it was a saving grace that everybody understood and spoke little Hindi in which we normally conversed besides English.

Initially, some administrative work was assigned to me. Mr. De was looking after the scientific and technical aspects of the observatory. This did not make me happy as all administrative work was finished in the forenoon and the entire afternoon was free for me. I started spending the afternoons in the library and the underground vault where seismographs and other instruments were installed. After a month of stay in a hotel, I shifted to CSO campus when Mr. Saha offered me the rear portion of his huge cottage that was lying vacant. My first bachelor's home was set up and furnished with necessary items for a comfortable stay. Staying in campus gave me a lot of extra time in the evenings. I was literally fascinated with the instrumentation and a variety of seismographs functioning in the underground vault, and often ended up spending evenings learning operation and maintenance of the seismographs. The vault was a very well-built structure 20 ft deep below the ground with a room in which a huge pier hosted the seismographs. The pier was isolated from the floor by a cavity 10 inches wide and filled with loose sand to avoid mechanical disturbances like walking disturbing the sensitive seismographs installed on the vault. I was fascinated to see the sensitivity of the seismometers that can pick up even the slightest movement of the ground. Those were the days of old seismometers where a sensor is suspended in a magnetic field and will respond to seismic waves due to relative motion of the sensor in the magnetic field generating a small current that will get recorded passing through a galvanometer.

One late evening while I was trying to fix a defective component of Wood Anderson Seismograph in the vault when I heard the sound of someone coming down the stairs of the underground vault. I turned back to see Mr. Saha standing behind me with a curious expression on his face. He was surprised what I was doing at this late hour in the evening. And when he saw me struggling in my efforts to fix the Wood Anderson seismograph's suspension wire, he sat down beside me and showed me how to do it in an efficient way. My commitment to work in my spare time even in late evening hours left an indelible mark on him and the very next day he issued an office order allocating scientific and technical duties to me. In those days, CSO was responsible to report any significant earthquake that occurred in the country and give a press release. There were 20 seismological stations all over India and with limited communication facilities it was a lengthy and tedious job as the data use to come via telegrams. The daily routine at CSO was to change the seismograms by 7 am and analyze them and if there is any significant earthquake then collate data from other stations and determine earthquake parameters to report to press.

My confidence grew leaps and bounds when Mr. Saha, who used to conduct a 4-week training course in Seismology for new recruits in IMD, chose me to take the classes. The nervousness of a young 21-year-old turned into excitement while teaching seismology theory and practice and made me pour through the copious notes that were prepared during my daily visits to library and my interactions with Mr. Saha which were extremely beneficial. These notes covered all the lectures spread over 30 h of lectures and afternoons for practical training on each and every aspect of seismology. Those four weeks I worked almost 15 h a day and realized my hidden strength to face up challenges that are thrown up unexpectedly and how you cope with them. This was a great learning experience and made me a very confident young adult. I was elated when I came to know that Mr. Saha used to listen to my lectures without my knowledge, sitting in the dark room next to the lecture hall, and admired them very much. After the course he became very close to me and introduced me to his family for a long-lasting relationship (Photo 1.7).

While this was going on I revived my interest in Badminton, got entry into Shillong Club and impressed members with my game. I was selected in the Assam Badminton Team for the All-India Inter-State Tournament. In those days, the entire north-east India was known as the Assam State. At the same time, there was the good news of my selection to Radio-Radar Training Course. But the destiny had other plans. On a rainy day while returning from club I slipped on a slope and fractured my ankle. This incident shattered my dream to represent Assam state in Badminton and miss the radar training course as my ankle took a while to heal. During the period of my recovery from the fracture, all residents of Meath Home poured their love and care and took turns to visit me. One of them was Mr. Tarun Kanti Dutta, a former CSO scientist, who had moved to Regional Research Laboratory in Jorhat and wanted to start a Geoscience Division, was staying at the CSO Meath Home campus with his family. Around that time the United States Geological Survey (USGS) had planned a global network of 120 seismological stations under a program named, "World-Wide Standard Seismograph Network (WWSSN)". In early 1960s, there were a variety of seismographs with different characteristics operating all over the world making it difficult to address scientific problems. To overcome this situation, WWSSN program planned to get global data from a network with similar instruments.

Photo 1.7 Mr. B. P. Saha (center), Mrs. Saha (next to Mr. Saha) and children

In India 5 locations were chosen under this program; Shillong was one among them with others at Kodaikanal, Kolkata, Pune, and Delhi. The USGS team was to arrive in a month's time, and I was still with my ankle in plaster. Fortunately, for me I recovered in time and was assigned the task of assisting the USGS team in the installation process. I dug deeply into the manuals and other literature on WWSSN equipment sent by USGS prior to their arrival. Re-location of the existing seismographs was carried out to accommodate Long- Period Press Ewing and Short-Period Benioff Seismographs on the same vault. Similarly, space was created in the recording room to accommodate recording drums of the new equipment.

The USGS team arrived CSO with all their equipment and were impressed with our site due to its low noise characteristics. The team was very happy with the arrangements, and it took two weeks to complete the installation. I vividly remember the excitement of seeing the first seismogram of a WWSSN station as it amazed me with the clarity of the recorded seismic phases, the overall quality of time marking and a beautiful calibration pulse indicating the health of the instruments, magnification etc. The USGS team stayed with us for a few more days and watched us operating the instruments and we at CSO felt happy to get hands on training under the guidance of experts and be a part of WWSSN network.

On 27 March 1964, the Alaska earthquake of M_w 9.3 occurred. This was the second largest earthquake of the twentieth century with over 40,000-time energy release in comparison to the Hiroshima or Nagasaki atom bombs. It was fascinating to look at the traces on our records at CSO observatory and realized that each wriggle on the record conveys its journey through the body of the Earth's interior before getting recorded. All that I had studied about different seismic waves flashed back into my memory and very little did I realized that I would be utterly swept away by this record that was inviting me to exploit innumerable opportunities it offers to people who have an eye for details and a highly imaginative mind to see through the working of planet Earth. Obviously, it was my destiny to be here in the abode of clouds in Shillong that is one of the most earthquake prone regions of the world. My first baby steps towards earthquake research were taken that day and started my journey to unravel mysteries of our mother Earth using the fascinating field of seismology. A modest beginning was made when my first research paper on *Direction of Approach of Short Period Microseism at Shillong* appeared in the Indian Journal of Meteorology and Geophysics in 1964.

My work at CSO observatory, analyzing seismograms, playing badminton and enjoying nature's beauty of Shillong was very fulfilling. An opportunity cropped up in the same year when a vacancy arose at the National Geophysical Research Institute (NGRI), Hyderabad, a laboratory under the Council of Scientific and Industrial Research (CSIR), New Delhi. It was an appropriate time to take a big forward leap in my profession. Working at a national research institute would offer much more facilities, provide opportunities and exposure necessary for a scientific career than at CSO which was essentially a service oriented organization with limited scope for research. Mr. Saha knowing my inclination towards research supported my decision and forwarded my application. There was a tragedy that struck that year when the first Prime Minister of independent India, Jawaharlal Nehru passed away. I recall that day when we were all glued to the radio for the running commentary of his funeral procession to Shantivan in New Delhi where his Samadhi was built.

On 21 September 1964 I attended an interview at NGRI, Hyderabad. Mr. M. S. Bhalla my senior at ISM was working at NGRI and offered me to stay at his home, which I did. Staying with his family was very soothing for me and calmed my nerves at the new place and also, he briefed me about the institute. In 1964, NGRI did not have its own building and the institute was housed at Type IV residential quarters of the sister Regional Research Laboratory, also belonging to CSIR. There were 20 candidates who were called for the interview for the 4 advertised positions, one of which being for a seismologist for which I was an applicant. My turn came at 12 noon. As I walked into the room, I could recognize Dr. Hari Narain, who was now the Director of NGRI. Other members were Dr. A. N. Tandon, who was then the Director, Seismology, IMD, stationed at Delhi; Prof J. Singh from ISM and three other experts whom I did not recognize. The interview continued for 30 min. Apparently, the committee members were pleased with my performance and by my single author publication. Dr. Narain informed me that NGRI is keen to initiate seismology related work at NGRI and wants to set up a modern seismological observatory at NGRI. He said they were looking for a person who can take up this responsibility from a scratch. As the interview ended, it appeared that I may have been selected.

Finally, the offer letter arrived in Shillong. In a short time of some 18 months in Shillong, a lot had happened. I had grown so fond of Meath Home and CSO staff. And above all the beautiful Shillong city. It was painful to leave. However, it had to be done. I joined NGRI at Hyderabad on 21 November 1964 and my baptism into seismology research was formally completed.

Life took several twists and turns with near death incidents. Fracturing my ankle, that looked so bad when it happened, turned out to be a boon in disguise, as it stopped my going for the Radio-Radar Training course that would have bound me to IMD for 3 years and an opportunity to join NGRI would have been lost.

Chapter 2
Scoring Firsts

2.1 Setting up WWSSN Type Seismological Observatory at NGRI, Hyderabad, India

My first day at NGRI was very eventful and explosive. I met Dr. Hari Narain, Director NGRI, in his office and had a very pleasant introduction. Dr. Narain told me that he wanted to set up a seismological observatory with specifications that of the World Wide Standard Stations Network (WWSSN) station at NGRI so that it can be a part of the global network. However, all his attempts with USGS have failed due to their prior commitments at other locations in India and they were not able to increase the number due to financial constraints. But Dr. Narain was determined to go ahead with his plan and had already obtained grants from CSIR to import all equipment from USA to set up this observatory on our own. Dr. Narain added that Dr. Amalendu Roy, Deputy Director of NGRI is looking after the issue of setting up of the seismological observatory and I should meet him. A room adjacent to Dr. Roy's office was allocated to me for operational convenience.

I met Dr. Roy in his office and was impressed by his elegance and precise way he spoke. After a brief introduction he told me that he has prepared a list of equipment for establishing a WWSSN station at NGRI in consultation with Dr. A. N. Tandon and handed over the list running into 3 pages. He asked me to have a look at the list. Since, I had the firsthand knowledge about the equipment having assisted the USGS team at CSO, still fresh in my memory, I glanced at the list and immediately pointed out the missing items and the shortcomings that needed to be addressed. This upset Dr. Roy a lot and he curtly told me:

> Mr. Gupta, I have prepared this list in consultation with Dr. A. N. Tandon spending more than a fortnight. I do not want it to be rejected by a green horn in 15 minutes. Take it with you and come back to me after a week.

This was not expected on my very first day at NGRI, within one hour of my joining. I was dejected and crest fallen. Dr. Mathur, who was sharing my room,

H. K. Gupta, *Harsh K. Gupta*, Springer Biographies,
https://doi.org/10.1007/978-3-031-63561-8_2

calmed me down and advised me to prepare a list and hand it over to Dr. Roy instead of verbal interaction with him. The very next day the revised list was ready with me, but I waited for a week as told by Dr. Roy. He was a globally respected geophysicist with his outstanding research papers in exploration geophysics. After his master's in physics, he had joined Geological Survey of India and was sent to Colorado School of Mines, USA for a MS in Exploration Geophysics. On his return, after serving GSI for several years, he had moved to IIT Kharagpur and started teaching geophysics. He had joined NGRI on an invitation from Dr. Hari Narain as Deputy Director along with other outstanding scientists like Drs. P. V. Sanker Narayan, S. Balakrishna, M. N. Qureshy, R. K. Verma and J. G. Negi.

Finally, a week passed, and I went to see Dr. Roy. As I entered his room, he asked me to take a seat and I gave him the list of the equipment that I had prepared. All the changes, and additions vis-à-vis Drs. Tandon and Roy list of equipment were clearly marked. He spent over 30 min, minutely going through the list that I had provided and the changes and additions that were suggested. He suddenly rose from his chair and asked me to accompany him to Director's Office. I had a sickening feeling while we entered Dr. Narain's office. After wishing him he said,

> You know, Mr. Gupta has joined a week back. I had given him a list of equipment to be procured for NGRI Seismological Observatory, which has been prepared in consultation with Dr. Tandon. However, he has prepared a list, that differs from our list. What should be done?

Dr. Narain in his characteristic cool style asked him to suggest a way out. Dr. Roy said,

> Under the circumstances, it is best to send both the lists to Mr. Leonard Murphy, in-charge of WWSSN at USGS for his comments.

The suggestion was accepted, and Dr. Roy wrote a letter to Leonard Murphy, who was heading the WWSSN project in USGS, USA, attaching both the lists, requesting him to comment at the earliest. Those days the fastest way of communication was through air-mail letters and the expected time of reply was about 20 days.

Fast-paced events over the last week, since joining NGRI, had totally exhausted me. I asked for one week leave as we must anyhow wait for a reply from WWSSN before we proceed further. Dr. Narain was totally aware of the situation and encouraged my taking a week off. Spending the time with family in our new home Madhu Jivan in Moradabad was refreshing and did calm down my nerves. Time flew by and I had to be back to Hyderabad after a week. The wait for the reply was agonizing and finally on 18 December 1964, at around 11 am there was a knock on the door of my office and Dr. Roy walked in and asked me to accompany him to Director's office. With great nervousness I pulled myself up and with my heart pumping as we entered Dr. Narain's office. Dr. Roy sat down on the chair opposite to Director and asked me to take the other seat. It sounded like an order and my heart sank in anticipation of what was to come as I took the chair. Without going into many details, he informed Dr. Narain that Leonard Murphy's response has come. I lost another heartbeat. Dr. Roy continued:

> Mr. Murphy has agreed with Mr. Gupta's list and has appreciated and complemented the thoroughness with which it has been prepared keeping the minutest details of the requirements.

With a firm expression on his face, he further addressed Dr. Narain and said with great admiration that:

> Under these circumstances it would be injustice to Mr. Gupta if he has to report to me after this incident. He is a very capable and confident young seismologist. I request you to make him an independent Project Leader for Seismology.

His words sounded like music to me. Dr. Roy was known to be very tough man but had a golden heart and appreciation for talent. I realized how sometimes appearances can be deceptive. The same day an office order was issued, declaring me a Project Leader, making me the youngest Project Leader in the institute at the age of 22 years. I was in 7th Heaven.

Without losing much time I embarked on my mission to establish the first seismological observatory on NGRI campus matching with the specification and standards of a WWSSN station. Once again, I reviewed the list of instruments and the accessories very minutely as these were to be imported from USA and any oversight might have delayed our task. After finalizing the list, orders were placed in mid-1965 and the delivery was expected after 2 years in May 1967. Since, we were on a shoestring budget, one important console of the setup could not be ordered. Instead of waiting for the additional funds that could delay the purchase process, it was decided that I shall develop the missing console at NGRI with the help of instrumentation people in the institute.

The next step was to select a suitable site on the campus to construct a vault for housing the instruments and recording system and other accessories underground. The main criterion was to see a place where the background noise would be minimum. It was a real challenge. Dr. Hari Narain told me to select a spot anywhere on the 150-acre plot allocated for the construction of the new NGRI campus in Habsiguda adjacent to the existing Regional Research Laboratory. Keeping in mind the future expansion of the institute and the city a location was selected at the farthest point from the proposed NGRI main building and the main road leading to the NGRI Campus, where hard granite rocks were exposed. I knew from my earlier experience at Shillong observatory that it must be away from all possible noise sources originating from anthropogenic activities. Dr. Narain was informed about the selected location. He invited Dr. A. N. Tandon for a second opinion.

One bright sunny morning, Dr. Hari Narain, and Dr. Tandon visited our newly selected site in the campus before a final decision was taken for observatory construction. Those days there were no roads inside the campus, and one must walk down from main road, a distance of about 700 m to reach the site. Dr. Tandon walked around the site and made a thorough inspection and approved of the site selected. Since, the granite outcrops near the surface were highly weathered the area needed to be dug up till the massive basement granite rock is reached from where the piers were to be raised and isolated with loose sand all around to cutoff high frequency cultural noise. At that time the only other structure on the NGRI campus was the magnetic observatory building in another corner. The civil engineering department of NGRI was roped in to prepare engineering designs of the seismological observatory and the seismic vault before the construction start. A vault design like that of the Shillong Observatory was finalized.

At the end of 1965 an advertisement appeared for a year-long training course at the International Institute of Seismology and Earthquake Engineering (IISEE), Tokyo, Japan that was to start from August 1966. For the selected trainees, the entire expenditure was to be borne by UNESCO, including travel, health insurance and other related expenses while in Japan. The requirement was a regular appointment in an institution, working in earthquake related research and a nomination by the Head of the institute. Several tens of nominations are generally received for this training course that is conducted at regular interval and only 15 trainees are finally selected for seismology. I knew that such a training course at beginning of my career would be a great opportunity to work with the best in the world. The competition was very tough but still the hope of selection made me think to go for it. Though there were other concerns as the construction of the observatory building was about to start, I decided to take a chance as the order of equipment were already placed and these were likely to be received just in time when the training course would finish. This issue was hesitantly brought to the notice of Dr. Narain who very gladly agreed to nominate me with a warning that there is no certainty of my selection in view of the stiff competition. I took a sigh of relief as the results of selection would be known in early 1966 and by that time, I would be able to complete the most crucial vault construction as other structures would be taken care by our engineering department.

In May 1966, a letter was received from UNESCO, informing me about my selection for the training and I was required to join by end of July 1966. By this time, the construction of the technically most crucial piers structure, that will host the seismometers, had been completed. The pier was very well isolated from the floor by providing a 9-inch cavity all around it to a depth of 8 feet and filled with loose sand providing almost complete isolation from the mechanical vibrations from people walking around or cars moving in the vicinity of the observatory. Dr. Narain was very pleased to see my selection letter. He stood up from his chair and hugged me and said that my selection is an honor for NGRI and he feels proud of me. Honestly speaking, I did not expect such a positive and encouraging response as he also promised that my concerns about the unfinished work will be taken care off (Photo 2.1).

2.1.1 Training in Japan

There was immense excitement as I boarded the flight from Hyderabad to Madras, from where the international flight to Bangkok on my way to Japan was to be taken. Those days there were no direct international flights operating from Hyderabad. This was my first international travel and anxiety kept me awake all through the travel despite excellent service and comfort of a bigger aircraft. I was picked at the

Photo 2.1 Dr. Hari Narain having a discussion with Harsh Gupta

Haneda airport and dropped at the International Student's Hostel where all trainees were accommodated. We were a batch of 22 trainees, 12 for the Seismology and 10 for the Earthquake Engineering courses. We had brief address by Prof S. Omote, the director of IISEE, Prof. T. Santo, Seismology Head and Mr. Watabe, Earthquake Engineering Head. After a weeklong Orientation Program, the regular course started. We were taught by great seismologists like Professors Hiroo Kanamori, Seiya Uyeda, S. Miyamura, K. Mogi, Yasuo Sato amongst the most famous ones. We also had lectures from well-known visiting professors such as Carl Kisslinger from USA and I. E. Gubin from Russia. The course was not difficult for me: thanks to the intense training given by Mr. Saha while working at the CSO. In my free time I tried learning Nihongo Language and shook my legs in dance classes. A wonderful opportunity came my way of meeting India's Ambassador to Japan, Mr. Badruddin Faiz Tyabji. The NHK (Nippon Hoso Kyokai) TV channel had contacted Indian Embassy for suggesting someone with good Hindi diction. My name was suggested by the Embassy to NHK. I worked for NHK for a while every alternate day in the evening. It was a very good experience and boosted my confidence.

Things were going on smoothly with the Seismology Course at IISEE. I learnt many new things. The Japanese professors were very happy with my effort to speak to them in Nihongo. Prof. Carl Kisslinger was particularly impressed with me and invited me to USA for a Ph. D. program with him. This offer I could not avail because of my commitments at NGRI. That winter we had a very good Christmas with friends from different parts of the world along with faculty members. On Saturdays I use to go Earthquake Research Institute (ERI), Tokyo University to analyze seismograms from WWSSN stations, which were available on micro-films. Prof. Sato made all arrangements for me to get these seismograms and I used to work till late in the evenings. He was very much impressed by my interest in seismology and agreed to work together on the dispersion of Love waves in Eurasia and that resulted in a publication in the Bulletin of Earthquake Research Institute. Within a few months I could communicate a manuscript that was published later. Meanwhile, I had the opportunity to attend the Republic Day function at Indian Embassy when our Finance Minister, Shri Morarji Desai visited Japan in January 1967.

The training course at IISEE finished in July 1967 and I bid farewell to IISEE. On my way to Haneda airport I had to visit Indian Embassy. When I left the embassy, there was heavy traffic jam and the flight was missed. The airlines were kind enough to accommodate me in the next flight to Manila. Since, my connecting flight from Manila was a day later, it did not affect me. I was glad to see Prof Yasuo Sato as he was on the same flight. After landing at Manila, Prof Sato bid me goodbye and as I reached my hotel room and switched on the TV. To my horror I saw the news that the flight I missed in Japan had a major problem while making a belly landing, as the landing gear of the aircraft did not function in Manila and several passengers had suffered severe injuries, including fractured limbs. Fortunately, there was no casualty, I felt a sigh of relief and thanked the traffic jam in Japan.

Back in NGRI I plunged myself deeply into establishing the observatory as the instruments have arrived and all necessary structure were completed. The console for which purchase order was not placed was rigged up with the help of our scientists at NGRI.

Day and night I spent long hours poring into voluminous installation and maintenance manuals that came with the equipment, brushing my copious notes that I had prepared at Shillong during the installation of WWSSN station at CSO Shillong. In the process there were quite a few hick-ups. I used to send long telegrams to Leonard Murphy, who still oversaw the WWSSN operations at USGS, seeking clarifications on certain issues and he used to oblige me with almost instantaneous responses. While using the manuals, a few errors were discovered that were pointed out to the manufacturers. They profusely thanked me and were surprised why these errors were not pointed out earlier. In the meantime, Indra Mohan and Balkrishna Rastogi were selected as Senior Scientific Assistants for the Seismology Program. Indra Mohan quickly took up sharing responsibilities at the seismological observatory, and I found him to be very competent. By the end of October 1967, the observatory was ready for operation. Dr. Narain and Dr. Roy spent over two hours with me and went through the minutest details and were satisfied. Since, this was the first time that a WWSSN system was being established without USGS help anywhere in the world, Dr. Narain wanted to make it a big occasion in the country. He decided to invite Dr. K. L. Rao, the then Union Minister of Irrigation and Power. While we were waiting for a confirmation from the Minister, a strong shaking on the early morning of 11 December 1967 woke me up. I realized that it was an earthquake, as I had similar experience while at CSO, Shillong as well as in Japan during my training course. There was no television in those days and I got the news from radio broadcast that a massive earthquake had occurred early in the morning at Koyna, a small township around 120 km away from Pune in Maharashtra. I rushed to NGRI and sought permission of the Director to start the observatory immediately to record the aftershocks of the Koyna earthquake.

So, the operation of the WWSSN type seismological observatory at NGRI began with Koyna earthquake of M 6.3 on 11 December 1967 before its formal inauguration. Later, we had a simple but very impressive ceremony of the inauguration on 20 April

Photo 2.2 Explaining the functioning of a seismograph to Mr. K. L. Rao (right) on the inauguration of the NGRI Seismological Observatory

1968. Dr. K. L. Rao, Union Minister of Irrigation and Power complimented NGRI to have established the first WWSSN equivalent seismic station anywhere in the world without the assistance of USGS. He spent quite a bit of time going through the entire recording system and understanding the minute details of the building, the seismic vault and instrumentation (Photo 2.2).

> Little, did I know that the Koyna earthquake will put India onto the world map and will serve as a field laboratory for RTS phenomenon and will attract worldwide attention. I realized later that destiny was guiding me to start a branch of seismology on Reservoir Triggered Earthquakes (RTS) in India which at that time was crystalizing and scientists were still apprehensive.

2.2 Discovery of a Double Crust Below the Himalaya and Tibet Plateau

My fascination with Rayleigh and Love surface waves, that began when I first looked at the Long Period WWSSN records at the Shillong observatory, continued even after joining NGRI in 1964. The romance with surface waves was to result in the discovery of a 65–70 km thick crust below the Himalaya and Tibet Plateau, which was published in the Bulletin of the Seismological Society of America in 1967. It is very gratifying to know that the discovery stood the test of time for over $5^{1/2}$ decades. Even in 2021, Dr. Gaochun Wang, a seismologist from China led a group of scientists from Europe, USA and Scandinavia and reported a 80 km thick crust below Tibet in the February 16, Issue of Nature. This finding is described here as the "*Discovery of a double crust below Himalaya and Tibet*" and explain the nomenclature.

The story began in Shillong while I sat in office wondering as to how such a breathtaking topography of Shillong and northeast India that includes the Himalaya in the north and Indo-Burmese Mountains in the east must have come into existence. Shillong, with its highest peak at about 2 km above the mean sea level, is a highland plateau with hill ranges all around yet resembling a rolling tableland. With depressions all around like Brahmaputra basin in the north, the Ganges delta in the west the monoclinal slope in the south leading into Bangladesh, Shillong looks like a *Roof* of northeast India. I always wondered how such roofs are sustained at great elevations. By this time average estimates of the layered Earth were known. The crust was about 35 km thick below the continents and 5 km thick below the oceans, mantle at a depth of 2900 km, outer core from a depth of 2900–5100 km and inner core from 5100 km to the center of the Earth at a depth of 6380 km. It was also known that continents ascribed to a laterally varying solid lithosphere that moves over a more fluid asthenosphere and explained the mechanism of *Plate Tectonics* theory.

After moving to NGRI in late 1964, and after completing initial commitment of setting up the NGRI observatory I had time to indulge into my fascination for earthquake records, especially looking at surface waves that travel all around the Earth multiple times generated by great earthquakes. The curiosity about mountain terrains led me to think and conceptualizes solutions in my mind about their formation. In a way I was using Albert Einstein's *Gedankenexperiment,* "Thought Experiment" approach of using conceptual rather than actual experiments when he proposed the theory of relativity. The question was, *"If the average thickness of crust is 35 km, how come high Himalaya and Tibet plateau sustain their heights".* Several researchers have proposed formation of the Himalaya to continent–continent collision process between India and Eurasia. This implied that the crustal thickness in such a scenario should be much more than the average thickness of 35 km. Elsewhere, researchers have suggested thickening of crust under collision zones and below certain mountains. For example, Gutenberg way back in 1933, from the study of longitudinal waves traveling through the Alps in Europe inferred the existence of roots. Byerly in 1938 inferred roots for Sierra Nevada in USA. Russian Seismologists had concluded in 1964 that in Eurasia there is considerable thickening of consolidated crust in the mountainous regions. Contrary to these suggestions, it was also observed that the crust beneath the Rocky Mountains is thinner than the Great Plains farther east, in the United States of America. But our knowledge of the thickness of the crust under the Himalaya and Tibet was limited, though some suggestion was made but no estimates were available till then. This may be due to lack of data from explosion seismology and lack of sufficient number of seismological observatories suitably situated for profiling.

This led to the thought of using Rayleigh and Love waves to address this problem. By 1964 the World-Wide Standard Seismograph Network (WWSSN) installed by the USGS had become fully operational and copies of records of any observatory could be obtained at a very little price for research purposes. While at Shillong I had seen records of an earthquake that occurred on 25 August 1964 in the Arctic region. The Rayleigh and Love surface waves generated by this earthquake had been very well recorded at Shillong. The entire path from the epicenter to Shillong was continental. I wanted to see how the surface waves generated by this earthquake looked on other seismic stations in India and the neighboring regions that have continental paths and then use the dispersion analysis to compute crustal thickness.

This issue was discussed with Dr. Narain at NGRI and his permission was sought to procure seismogram of a few WWSSN stations to purse my thought process and solve the mystery of the high Himalaya and Tibetan Plateau. He immediately agreed to my request as he saw merit in my arguments. An order was placed to obtain seismograms of Quetta, Lahore, Chiengmai, Hong Kong and Seoul. Seismograms from Shillong and Delhi were also obtained. The seismograms arrived sometimes towards the end of 1965. The idea was simple. Waves recorded at Shillong, New Delhi, Lahore, and Quetta pass through the high mountain regions of the Himalaya and Tibet Plateau while those recorded at Seoul, Chiengmai and Hong Kong do not pass through these high mountain regions. An in-depth analysis of the dispersion of Rayleigh and Love waves of all the 7 WWSSN stations seismograms were compared with the theoretical dispersion curves computed for a three-layer earth model revealed a thick crust of 65–70 km below Himalaya and Tibet. As the results were startling, we reported in a communication to the Bulletin of Seismological Society of America (BSSA) before I left for my training course at IISEE in Japan in 1966.

I vividly remember that day in Japan when I came to know about the publication of this landmark paper in BSSA. Prof Carl Kisslinger who was one of our teachers at IISEE training course, surprised me as he broke this news in the class to a great applause from everybody. The same afternoon Prof Sato who was an authority on surface waves studies and known globally for his path breaking research on surface waves complimented me on my work. He said he will forgo his lecture and asked me to explain my research and results in the BSSA to the benefit of all the other trainees. It was a rare opportunity and a recognition from Prof Sato, as he was known to be a very serious professor, who spoke very few words and being very strict with his students. More surprise was in store. As our class got over, Dr. Omote announced a get together to celebrate the occasion. While addressing the gathering of all staff and students of IISEE he specifically made a mention that how fortunate they were to have a participant from India who at the young age of 25 published a landmark discovery in one of the most reputed international journals. He highlighted the fact that the presence of an extremely thick crust below Himalaya and Tibet Plateau region is reported for the first time and reiterated that it was a path breaking research with limited data sets but backed by a sound conceptual thought process. This was a big story that was acclaimed internationally because such an estimate of a double crust was not reported from any other region in the world, till then.

Since, the average continental crustal thickness is about 35 km I call the discovery of a 65–70 km crustal thickness a *Double Crust* below the Himalaya and Tibetan plateau. In a recent communication in June 2022 the Russian Academy of Sciences acknowledged my 5 decades old discovery while making me a Foreign Member of the Academy.

2.3 Common Characteristics of Triggered Earthquakes

Two events, the establishment of the seismological observatory at NGRI and occurrence of Koyna earthquake in western India in December 1967 laid the foundation of the Reservoir Triggered Seismicity (RTS) studies in India. It was a golden opportunity to study field effects of the earthquake in and around the epicentral region and try to make out some sense in combination with the earthquake data from NGRI observatory. On my initiative Dr. Narain approved my plan to visit earthquake affected region in Koyna and even accompanied me to the field. In Koyna we met Dr. Banerjee of Central Water and Power Research Station (CWPRS) who was in-charge of a seismic network around the Koyna dam that was constructed for hydro-power generation. He briefed us about the Koyna Dam where impounding of the reservoir began in 1962. Following the initial impounding in stages as and when the dam height was raised, there were reports of felt earthquakes in the region. People were concerned because the region never experienced such earthquakes, earlier. To record these tremors, CWPRS had set up a network of 6 stand-alone seismic stations, around the dam covering the entire reservoir area. The network recorded several small tremors frequently, the strongest of them was on 13 September 1967 with a magnitude of M 5.7 followed shortly by the M 6.3 December event. We saw several damaged houses in Koyna Nagar, a locality occupied by the staff members working in the Koyna Hydro-electric Project. I spent a lot of time in making minute observations of the displacement caused by the earthquake. Two very interesting observation were made, one of the popping up of tiles against each other on the span of the Koyna Dam indicating compressive forces on the dam structure and the other was of rotational movement of mile- stone pillar. This earthquake claimed over 200 human lives in the region. Our observatory in NGRI recorded all the aftershocks of the Koyna earthquake. Back in Hyderabad we procured 6 components WWSSN records of several seismic stations having a very good azimuthal coverage. There was some confusion about the magnitude given by International Seismological Summary and India Meteorological Department. We did a very systematic study of the seismograms from all azimuths and calculated the magnitude as M 6.3. This was the first paper published in Nature in 1968, from NGRI (Photo 2.3).

The near vicinity of earthquake epicenter near the Koyna dam site and absence of any earthquakes prior to this event made me think of a possible relationship with reservoir impounding. But the real trigger to seriously pursue Koyna earthquakes came after I attended a meeting on Koyna earthquake at the Bhabha Atomic Research Center (BARC) Bombay. The government of India was concerned about this earthquake because it occurred immediately after the reservoir impounded to its full level and the closeness of its epicenter to Mumbai, the financial capital of India. Several senior scientists and researchers from IMD, CWPRS, BARC and IITs were invited to deliberate and discuss this earthquake. The meeting was chaired by Dr. Vikram Sarabhai, a space scientist of international repute. He was Chairman of the Indian National Committee for Space Research (INCOSPAR) that later became

Photo 2.3 Rotational displacement of a pillar at Donachiwadi observed during the field study of the 10 December 1967 Koyna earthquake. The earthquake occurred at around 4.31 am, Indian standard time on the 11 December. According to Universal Time, it occurred at 22.51 h on 10 December. So, it is known to be the 10 December 1967 Koyna earthquake

Indian Space Research Organization (ISRO). The agenda of the meeting was to ascertain, whether the reservoir impounding was responsible for Koyna earthquake or not. The opinion was divided amongst the scientists with points and counterpoints being discussed in an animated manner. Suddenly, Dr. Sarabhai pointed a finger at me, sitting in the 10th row in the hall and asked, *Harsh, do you have a view to express?* This sudden and direct question surprised 200+ scientists present in the hall, and all eyes turned toward me. For a second, I almost froze as I did not expect Dr. Sarabhai to remember my first name and identity after months of our meeting during his very brief visit of few minutes to NGRI observatory during establishment. Immediately, I regained my posture and told him in a confident voice that the work on this earthquake is already underway and a definite conclusion about its association to reservoir or not would be known, soon. I could see that everyone present there took notice of a confident 25-year-old novice who had Dr. Sarabhai faith in solving this important issue.

This meeting defined my path for the next three decade of research in the field of Reservoir Triggered Seismicity (RTS) as a commitment was made to find an answer. With missionary zeal the work started with studying the seismograms of Koyna earthquakes that were procured from several other stations along different azimuths. We showed that the mainshock of M 6.3 Koyna earthquake was a multiple seismic event and with a very precise study of the travel time of different seismic phases the entire rupture history in the source zone was mapped. Our results found acceptance in a paper published in BSSA in 1971. We started looking deeper into other RTS cases worldwide. Slowly patterns started emerging that led to the identification of several common characteristics that differentiate triggered earthquakes with normal tectonic earthquakes in a region. To understand and appreciate the results it is important to know the nomenclature of earthquake sequences. A mainshock is defined as the largest magnitude in an earthquake sequence in which smaller magnitudes aftershocks follow till the seismic activity stops, completely. The foreshocks occur prior to the mainshock and are smaller in magnitudes but are tricky to define and are identified later when the sequence ends. The length and distance of the causative fault often limits the definition of foreshocks and aftershocks. Three types of patterns were identified by Kiyo Mogi, a Japanese scientist in early 1960s. In Type I pattern a mainshock occurs without any foreshocks followed by several smaller magnitude aftershocks. In some rare case a few isolated foreshocks are reported. In Type II pattern the mainshock will have several foreshocks and will be followed by several aftershocks and in Type III several microearthquakes occur in a localized area without any distinguishable foreshocks, mainshock and the aftershocks pattern. These are called swarm activity, often observed in the Stable Continental Regions (SCR) of the plate interior.

These patterns reflect mechanical properties of the medium defined by the grade of heterogeneity from a homogeneous state. From our studies in the Koyna region and from other RTS sequences, globally, a few common characteristics were identified that differentiate triggered earthquakes with that of normal tectonic earthquakes in the region. We suggested that RTS sequences are of Type II Mogi Model, the ratio of the largest aftershock to the mainshock will be higher, the decay of aftershocks will be slow and follow Omri's Law and the 'b' value will be higher than that of the regional 'b' values. The 'b' value is the slope of a relationship between the number of earthquakes and the magnitude as defined by Gutenberg and Richter. Since, this was a new finding, there was initial reluctance to accept our results. A few of our papers on Koyna earthquakes were published in BSSA in 1972 that slowly brought global acceptance and attention. When 1993 Latur earthquake of M 6.3 occurred near Makini Dam, 300 km northeast of the Koyna, the RTS common characteristics were found to be very useful to rule out the role of Makini Dam in triggering this earthquake.

As our research on Koyna earthquake was gaining global attention, I received an invitation letter from IISEE, Tokyo asking me to apply for a forthcoming advanced UNESCO seismology course starting in 1971. To grow in a profession and make your research at the international stage it is very important to go global. When I approached Dr. Narain with the letter, he said very encouragingly that my first training course in 1966 has helped NGRI to develop a strong group in Seismology research and the establishment of the observatory in the campus. He felt that this advanced course will greatly help me and the institute. I left for Japan once again to attend the UNESCO advanced course in seismology after I received my selection letter in April 1971.

In Japan I met Prof Mark Landisman from the University of Texas at Dallas (UTD), USA. He was one of the visiting professors at IISEE and a good friend of Prof Sato. Four of our papers on RTS appeared in BSSA and Geophysical Journal of Royal Astronomical Society at that time. Those days research institutes and universities use to get the hard copies of scientific journals and these journals were available in Prof. Santo's office, who was now the Director of IISEE and my work in seismology was known to him by this time. It was a great moment of satisfaction when Prof. Santo hosted a get-together to felicitate me during the training course. The get together was attended by Prof. Sato with whom I worked earlier and Prof Landisman. Later, in the evening Prof Landisman came to my room and discussed my work in detail and was greatly impressed with my efforts. He invited me to UTD, USA for a short visit to write a proposal for funding to National Science Foundation of USA for a joint work with him. This was agreed to and I got all the administrative clearance with the help of Dr. Narain. It was a memorable moment when I met Prof. Kiyo Mogi whose work had been extensively used in studying triggered earthquakes. He complimented my work and was thrilled to know that I am putting these ideas in the form of a book entitled 'Dams and Earthquakes' that was to be published by Elsevier Publishers.

A brief visit was made to UTD, USA after finishing the training course in Japan and a proposal was written and submitted for a joint work with Prof. Landisman. On my way back to Hyderabad from Dallas, I met Ms. Frances Delany, who was looking after the administration of the Geodynamics Program in Paris. I chanced to see an article by Prof J. P. Rothe in Paris, France entitled, "Fill a lake, start an earthquake" and was very much impressed and thought of meeting him at the first opportunity. I came back to India at the end of 1972. In 1973 I got an invite from the Royal Society in London for a meeting on the discussions on RTS. Around that time one of our papers on earthquake magnitudes and their relationship with source that was published in Geophysical Journal of Royal Astronomical Society was adopted as a chapter in the Open University textbook for a second level course in Geophysics. This was one of the great recognitions of our work in seismology at the global level. During the meeting in London, I had the opportunity to meet Prof Rothe who was aware of our work on RTS, and we had intense discussions on our research. I requested him to write the Foreword of our book that was in the finishing stage. He agreed to my request.

Back in India I got busy in giving finishing touches to our book with Dr. B. K. Rastogi as a co-author when Prof. Prof Landisman informed that our proposal to NSF was approved and he was looking forward to starting work with me at the UTD, USA. It was a good opportunity to continue my research work as well as try my hand at teaching in the university. Before leaving for USA, I finished the manuscript of our book and sent to Elsevier for publication. While I started our work at the UTD, our book 'Dams and Earthquakes' was published in 1975 when I completed 33 years of my age. There was a coverage on the front page of Dallas Morning News, with my photograph, with seismogram spread in front of me. Mr. Cecil Green a British-born American Geophysicist, who helped UTD and was one of the founders of 'Texas Instrument', a top-class company in the US invited me and Manju for tea at his home. It was indeed a very pleasant visit, and I presented him a copy of the book, 'Dams and Earthquakes'. He and his wife Iida Green walked us back to our car.

Our book became very popular and due to some requests from international community it was translated into Russian and Chinese languages (Photo 2.4).

While at UTD, in 1970's, there was a lot of attention on Geothermal Resources. I submitted a project to NSF with Jim Combs on the utilization of earthquake generated seismic waves to quantify geothermal resources. The project got funded. At that time, I read a lot and wrote notes while self-learning about geothermal resources. Mr. Van Eyesingha of Elsevier, with whom I had been in contact for the book 'Dams and Earthquakes', was on a visit to Dallas and came to see me in my office at UTD. Sitting across the table he saw the copious notes related to Geothermal Energy. He took my permission to glance through them. After spending some time, he suggested

Photo 2.4 From the right: Me, Prof. Mark Landisman, Mrs. Celia Landisman, son Andrew, and front: daughters, Jessica and Carole

that the basic work that I have done on Geothermal Energy could lead to publication of a book on the fast-emerging field of Geothermal Resources. The offer was indeed interesting, and I sought time to think about it. Finally, the suggestion was agreed to, and a proposal was developed and submitted to Elsevier about writing a book on Geothermal Energy, which was accepted.

On the social front there use to be get-togethers of the Indian community. In one of our meetings an issue was raised that our children being taught in American Schools were losing touch with India and Indian culture. It converged that ideally, we should have weekend classes for a couple of hours, where someone could stand and tell children about Indian culture. Mr. Seekri, being among the senior Indian residents took up the responsibility to find a suitable place for such classes and organize someone to take responsibility to talk to the children. As we lived in the same apartment complex, we sat down on a Saturday and decided to approach the Richardson Library, which was very centrally located and had the facility of conducting the proposed class for some 20 children. It was also decided that 4 of us, that included me and Seekri plus two other friends, would each take a class for one weekend each and see how it develops after a month and then decide the future course of action. We next approached the Richardson Library, and they were more than happy to lend us an excellent lecture room from 9.30 am to 00.30 pm. The idea was that we will conduct the class from 10 to 11 am, then have a 15-min gap, offer some snacks to children, the class will then resume at 11.15 and end at 12.15. In the next 15 min we will clean up the place. The announcement of this program was made, and we had an enrolment of 26 children. On the first weekend, the class was taken by me. Parents of all the children came. They brought so much stuff to eat that in the break, we invited all those present to join in snacking. Being the first day, I mostly spent time in learning all about the people that I was to handle. I kept talking to them in Hindi or English, often repeating myself so that all understood what was being told.

Everyone was quite happy. Children enjoyed it a lot, as there was no text, just talking and learning about themselves. For children it was a rare opportunity to meet so many others of Indian origin. After the class all the parents helped in cleaning up the place and we thanked the Richardson Library Officials. Next week it was Seekri's turn. However, a day before the weekend he fell sick. No one else was available so he requested me to stand for him. With a little hesitation I accepted to stand in his place and enjoyed talking and interacting with the children. On the third weekend, the gentleman who was to take the class was sent for a business meeting to Honolulu by his company, so I stood for him, too. By this time, I had known all the children by their first name and had enjoyed interacting with them. They also appeared to be happy with me. On the 4th weekend, I went to the Richardson Library. As soon as I entered the class all children clapped. I introduced the new teacher. He was finding it difficult to communicate with them. In the break, I requested him to permit me to take the second half of the class after the break. He was only too happy. By now I had realized that how much I had started enjoying interacting with the children. From then onwards, it became a routine for me to be at the Richardson Library every Saturday and Sunday. This resulted in my earning the title 'Guru' from the Indian community. Even today in 2023, whenever I go to Dallas (usually in December after attending the AGU Fall meeting) and meet my old friends they address me as 'Guru'.

A very interesting incident happened during my stay in USA when I received an invitation to give a talk on RTS in a two-day seminar at Caracas, the capital of Venezuela. It was indeed a prestigious invitation and an opportunity to further expand our footprints in the South American continent. My talk at the seminar was very well received that led to lot of discussion amongst the participants. After my talk, while I was in my seat in the front row one of the organizers came and handed over an elegant and ornamented name card with the request to join the sender for a dinner that evening. I obliged and accepted the invitation and sent a thank you note to him. After the session I met the host who was not known to me earlier. He spoke to me in Spanish and immediately shifted to English when he realized that I don't understand the language. He said that he listened to my talk with great interest and was impressed particularly on the stress for the need of proper assessment of induced earthquakes related to oil production. He mistook me to be from Venezuela and felt proud that a young Venezuelan has delivered an invited talk and he wanted to honor and encourage me and hence, the invitation. Realizing the mix up I offered to excuse myself from the dinner to which he did not agree and insisted that I should enjoy the hospitality of Venezuela to which I relented. At 7 pm I was picked up from hotel in a limousine that took me to a small private airport, where a helicopter was waiting for my arrival. After a 15-min ride in the chopper we landed on a small Caribbean island owned by the host. Another limousine took me to his sprawling 6 storied mansion. My host received me on the 4th floor that he occupied. We chatted for about half-an-hour when a bell rang at 8:30 pm. He explained to me that all the flats in the building are occupied by his family members and every day this bell is a reminder to assemble for dinner. I was introduced to the family members and we had a lavish dinner. At 10.30 pm, it was time to say good- bye. The mode of return journey was the same as the one while coming. Back in my room I reflected on the happenings on that day and that visit remain etched in my memory.

Time was flying by and I was busy with my work at UTD when I got an invite from the University of North Caroline at Raleigh to give a talk in a meeting on geological aspects of the Indian shield. I was thrilled to know that Dr. Narain and Dr. Naqvi along with Dr. B. P. Radhakrishna were also attending. After the meeting I invited Dr. Narain to Dallas and stay with us. Dr. Narain met Drs. Landisman and Combs at UTD and took great interest in our work. In between I have to be away for a couple of days and my family had good time with Dr. Narain. We went to airport when Dr. Narain was returning back. Before departing he told Manju, *please remember what we discussed* with a mysterious smile and left me with a puzzled look.

In 1977 a senior scientist position was advertised at NGRI that suited my background. Suddenly, I was struck by a flash and understood Dr. Narain's mysterious smile and his parting words to Manju while returning back to India. Sensing his desire, though conveyed indirectly I thought that it is time to go back to India and develop seismology research in the country. I was overwhelmed to receive the appointment letter, and it was a welcome home coming. Back in India, I got busy with my work and revived several projects that needed my attention. In between there were extensive travels to China, France, UTD and several other countries to attend seminars and other invitations. The work on geothermal studies was revived and I got busy in completing my book. A book entitled 'Geothermal Resources: An Energy Alternative' was published in 1980 by Elsevier Scientific Publishing Company. In between this period several research papers were published in reputed international journals and work continued with several students at NGRI.

The rich and varied experience I gained while working in different countries on either assignments as visiting faculty or in seminars and global projects laid the foundation and prepared me to take up several challenging tasks in the country as and when called for. It was great time establishing institutions and taking leadership roles that benefited the country. A new phase of life was about to start, and a most challenging task was awaiting to be taken up.

2.4 Establishing India's First Permanent Station in Antarctica

It all began when an offer was made to join the Centre for Earth Science Studies (CESS) in Trivandrum (Thiruvananthapuram, now) in Kerala as the Director in 1982. That was a great year for me as I was awarded the most prestigious Bhatnagar Prize given to scientists below 45 years of age. Everything was moving smoothly for me in the profession with several foreign travels in connection with International Geodynamics Project and I also became an Executive Committee member of the International Association of Seismology and Physics of the Earth's Interior (IASPEI).

A major task before me was to plan a campus for CESS that was operating from half a dozen private houses in Trivandrum. Obviously, the performance of the scientists at the institute was not satisfactory despite having a great talent pool. The facilities and an environment required for scientific research was lacking. It was necessary to transform CESS into a world-class research institution with state-of-the art laboratories, computational facilities, a modern library, auditorium, and other supporting infrastructure in a good environment. My efforts with the Chief Minister of Kerala Mr. C. Karunakaran succeeded when a 19 acres plot at Aakkulum in the outskirts of Trivandrum and allocated. Without wasting much time, the planning for a new campus was started in June 1983, little realizing that in between I may be called upon to fulfill a major Indian dream of establishing the first permanent station in Antarctica.

In 1983, the Department of Ocean Development (DOD), New Delhi, called for project proposals for the scientific work to be undertaken during the proposed 3rd Indian Scientific Expedition to Antarctica. I submitted a proposal for setting up a network of 5 seismic stations to record Antarctic earthquakes. The proposal was appreciated by the DOD Committee but manpower to run 5 stations could not be provided. Dr. S. Z. Qasim, Secretary DOD instead offered me to lead the expedition and establish the first permanent station in Antarctica where scientists could stay round the year for wintering. He told me that Mrs. Indira Gandhi, the Prime Minister of India was very keen to have Indian presence on the icy continent throughout the year. I had just completed 40 years, then. It was a great opportunity, but as the work had already begun and planning of the CESS campus was underway, there was the need to have discussion with the Kerala Government authorities and my fellow scientists before a decision could be made. After intense discussions everybody felt that since there will be time required for obtaining several clearances, and the required funds were still to be allocated, before the construction of the campus starts, there is no harm in taking up this assignment. Everybody was excited that their Director will be leading the 3rd Antarctic expedition that will bring glory to the Centre and they assured that the administrative and civil engineering issues will be taken care by them. Even the government of Kerala felt that it would be a great honor for the state when I met the Chief Minister. With such positive support and backing of the Kerala government I decided to take up the challenge that would put India in the elite group of Antarctic club. Before embarking on the mission to Antarctica, a blueprint of the plan that had been prepared for CESS campus with a promise to complete the task after my return, was given to the Officer-in Charge of CESS during my absence.

2.4.1 Preparations for the Expedition

We had a meeting with Prime Minister Indira Gandhi where Dr. Qasim introduced me to her. She was very happy to see a young 40-year-old scientist chosen to be the Leader of the Expedition. The meeting was very interesting. She asked several questions for which I had no answers. She told me that I should call on her again in a week's time. At that time, I contacted the UK High Commission and the Soviet Embassy to elicit as much information as possible as both these countries had permanent stations in

Antarctica. The discussions were very fruitful and the information, startling. It was made clear to me that no country had so far succeeded in setting up a permanent station in Antarctica in one Antarctic summer. The chances of success were hardly 10–15%. One very important advice that proved critical was the cargo loading sequence in the ship's hold, first ones to be unloaded in Antarctica is to be put on top of the hold and so on. Examples were provided when the entire expedition was called off as things required first were at the bottom of the hold and the entire hold cannot be unloaded on the prevalent fast ice in the month of December on the periphery of Antarctica. The other advice was about the precautions while unloading the Piston Bulley on the fast ice with adequate care so that they do not sink because of the lack of strength of fast ice. An example was cited, where the Piston Bully lowered on to the fast ice, without checking its strength, just sank and it could not be retrieved.

As desired, I met the Prime Minister after a week, brimming with confidence and loaded with all the information required to responds to questions. She was happy with the efforts made and told me to keep meeting her intermittently.

We then, got down to the formation of a scientific team with different expertise and finally a 16-member team was finalized. Two women scientists Aditi Pant, a biologist from the National Institute of Oceanography (NIO) and Sudipta Sengupta, a geologist from the Jadavpur University were part of the team. Other scientists included were S. W. A. Naqvi, an oceanographer from NIO, M. R. Nayak and M. Manoharan, Communication Engineers from NIO, A. K. Hanjura from National Physical Laboratory, R. K. Singh a field geologist from Geological Survey of India, K. J. Mathew, a marine biologist from Cochin, Madan Lal, a geologist from the Oil and Natural Gas Commission (ONGC), A. K. Bakshi, a chemist from Delhi University, L. S. Rathore, a meteorologist from India Meteorological Department, Ashutosh Singh, a HAM operator and Jitu Vartak, a photographer from the Films Division.

The responsibility of the construction of the Permanent Base Station rested with the team from the Corps of Engineers of Indian Army headed by Major Pawan Nair. Other officers included Captain H. C. Lohumi, Captain Anil Harnal, Captain Ajit Patil, and Captain A. T. Parnaik. It was the norm for other countries as well, to entrust the task of setting up of permanent bases in Antarctica, to men in uniform. There were 12 members from Indian Navy for providing logistic support with two Chetak Helicopters. The Naval Officers were Cdrs. V. P. Sathiamoorthy, and S. K. Chandna, Lt. Cdrs. R. S. Gill and A. S. Chouhan, and Lt. M. S. Khela and A. A. Khan. The Indian Air Force lent logistic support by providing two MI 8 Helicopters and a team of 12 officers and men including Wing Cdrs. R. S. Tandon, B. D. Madhok, and V. Natarajan, Flt. Lt. Ajit Kumar, V. M. Khanna and N. M. Rai and Sq. Leader. K. I. Trivedi. There was also a meteorologist from Navy, Lt. Cdr. N. P. Singh. The medical team consisted of Sq. Ld. S. K. S. Puri and Major B. K. Singh.

The decision was that in case we succeed in setting up the Permanent Base in this expedition, it was to be manned by a 12-member team headed by Lt. Col. S. S. Sharma along with S. R. H. Rizvi, Mr. Prabhu Matondkar, Captains Paramjit Singh, R. R. Sinha, and Ram Kumar, Surgeon Lt. Cdr. A. Banerjee, Nayak Subedars S. Joseph and V. S. Rana, Hav. Padmanabhan, and Lance Nayak S. Sadekar and C.

S. Thampi. The entire expedition team members underwent a week-long training at Machoi Glacier in the Himalaya at an average elevation of 4800 m with its highest peak at 5448 m height. We were very excited during the training and were raring to go and do something great for the nation. After the training we had a very brief family time before embarking on a mission which was to be four months for some and 16 months for a few members in the testing conditions with six months of darkness during winter.

2.4.2 *The Journey*

The ship Finnpolaris, from Finland was chartered for the expedition. It arrived Goa on the 25 November. Dr. Qasim had fixed 1st December 1983 as the date of departure. I visited the ship with Dr. Qasim and was introduced to the Captain of the Ship, Lasse Kulju. Dr. Qasim told him to follow my instructions on all matters connected with the Expedition. In the evening, I met Major Nair, a bright young Indian Army officer responsible for the construction. I passed on instructions on the loading sequence in the hold of the ship as well as precautions while lowering the Piston Bulley on the fast ice. Pavan had a good idea about what all was in the 80 feet deep hold of the Finnpolaris but did not know the sequence of requirement after arrival in Antarctica. It was agreed to rearrange the loading and this took about two days and delayed our departure. We set sail on 3 December 1983 (Photo 2.5).

The atmosphere was so electrifying and the excitement palpable as we waived, standing on the deck, to our near and dear who had come to give us a very warm send off and good wishes for the success of the expedition. We all kept on waiving till their figures dissolved in background of the enormity of restless sea waves. Several thoughts crossed my mind, and I knew a new dawn is waiting for us the next day.

Photo 2.5 Helicopter being lowered into the hold of the ship Finnpolaris

That day we had our dinner at 9 pm in two shifts because the dining hall could only accommodate one-half of our 81 expedition group members. As the Leader of expedition I had a separate cabin with all amenities for personal care and furniture and space to use as my office. After the dinner we all retired by 11 pm. All members were briefed about the daily routine to be followed till we reached the destination and duties were assigned to every individual as agreed and decided prior to boarding the ship. A typical day on the ship would start with technical meetings with the construction and logistic group mostly of army and defense personnel followed by interactions with scientists, most of them very young and budding, carrying great dreams in their eyes to excel in whatever they had planned during their short and long duration stays. There was a great excitement when the first scientific experiment of the expedition was launched next day on 4 December 1983 when a weather balloon was released into the atmosphere from the ship's deck.

At 04.02 pm on 6 December 1983, the ship crossed the equator, and a very colorful ceremony took place on the deck. Several of the 3rd Expedition Team Members crossed the Equator for the first time and we had the '*Crossing the Equator Ceremony*' and blessings from Lord Neptune. King Neptune permits the ones crossing equator for the first time after giving them a tough time through the veterans who have had crossed equator many times. There was gaiety and fun spread all over the deck of the ship. The sea veterans had dressed up and King Neptune's court was set up. The first-time equator crossing members were put to various naughty pranks. The most common was to ask the first time crosser whether he or she could see the 'Southern Cross'. And to help one see the 'Southern Cross', a binocular made of two Coca Cola bottles wrapped together, filled with sea water was provided. So, when you look up through this improvised binocular, your eyes get filled with salt water. The fun continued for a couple of hours, then 'Equator Shower' was given to all through the pump of the ship. In the end, a certificate was issued for having safe journey across the equator by the Captain of the Ship.

Proper planning is utmost important when one is involved with mega projects like building of India's first permanents station in Antarctica. Appropriate timelines need to be built and stuck to, keeping some margin for un-expected happenings. Managing a group of 80+ personnel from army and civil backgrounds required skills *extraordinaire* to succeed. I was happy that the team was highly motivated to succeed in the mission.

It was well known that the sea around Antarctica is frozen, and the fast ice starts disappearing in the Antarctic summer month of December. January and February are two months with minimum frozen sea around Antarctica. These are the two months when ship can reach the shelf ice and unload the material usually carried on sledges. It was therefore essential to finish all construction related activities by the end of February and avoid the most undesirable situation of getting ice locked. It was also known that in the region, where we were planning to build the station, usually has white outs and blizzards lasting up to 25 days during the months of January and February. A whiteout is a situation created when strong winds in Antarctica carry a lot of fine ice particles in air, which under multiple reflection of the sun make the atmosphere so dazzling that it makes you feel blind and you cannot see anything. It was therefore necessary to work out a plan to complete all the construction work in 35 days.

In one of our daily meetings, while discussing the construction activities and scientific activities to be undertaken in Antarctica, 3 of our team doctors raised the concern about the inadequacy of the medical facilities on the ship, that could take care of all the possible emergency scenarios. They informed that all the necessary medical equipment meant to be deployed at our Wintering Base Station are available and needed an exclusive room on the ship. Considering the importance of the request a room was allotted to them by rearranging some of the scientific equipment appropriately.

In the 35-day working-time window available to us, we had to perform several tasks like, (i) select a site for station, (ii) dig into ice down to six feet and place the raft foundation for the structure, (iii) assemble the structure, (iv) provide iron sheet cladding outside and an asbestos sheet lining inside, (v) furnishing, setting up generators, ice melting tank, (vi) anchoring the structure to ground to secure the structure against the strong winds, (vii) demonstrating the worthiness and the integrity of the structure by staying in them for a couple of days, (viii) to make sure all internal fittings function normally so that the wintering team who are going to stay in them are safe and secure for the next 1 year when the ship leaves them in March till the next team arrives in late December.

Initially, a week was planned for site selection but as a possible site location was already recommended by the earlier expedition based on the regional maps prepared by the Norse Polar Institute, it was decided to cut down the time for site selection to just one day, or preferably half-a day.

Our route to Antarctica was via Mauritius where we had our stop for 4 days after sailing for nearly 4500 km. It was important that we refill our requirement of fresh food to last for some more time before we shift to the frozen stuff. Also, there were some technical issues related to winches and needed attention to fix it. There was also some malfunctioning of refrigeration system of the food container that needed repair. After fully satisfying ourselves, we sailed off from Port Louis in Mauritius on 14 December 1983 in the afternoon. From Goa to Mauritius the journey was quite comfortable. However, after leaving the shores of Mauritius as we moved further south, we started feeling the greater swing of the ship. As we neared 40° South Latitude the sea was very rough that increased the rolling and pitching of the ship making the first timers a little nervous and sick as the intensity kept increasing. While, quite a few expedition members felt seasickness I was quite normal due to my earlier experience while working on high seas around Japan and India. Things were no different this time as well. Sometimes, I wonder to have missed the rare experience of getting seasick. To counter the extreme rolling and pitching of the Finnpolaris, the speed was reduced to 5 knots or less for considerable time till we crossed the *roaring south*, the region around 40° South on 18 December 1983. It was 4 am on 23 December, I got up early and had the first glimpse of an iceberg. The location was checked. It was 55.33°S and 32.03°E. I made a rough mental calculation, and suddenly, a shiver of excitement went down my spine as I realized that very soon, we would be entering the floe ice regime of Antarctica where the sea is half frozen.

Photo 2.6 Penguins welcoming Finnpolaris arrival in Antarctica

Floe ice is a cohesive sheet of ice floating in the water generally seen around Arctic and Antarctic regions. In Polar Regions water occurs predominantly in its frozen state. It either falls as snow to contribute to the growth of ice sheets and glaciers or it drifts on the seas as ice floes. Our ship encountered the first floe ice at 65°S Latitude on 24 December, and it was interesting to watch the professionalism of the Captain Lasse Kulju who maneuvered the ship very skillfully and with great calmness and confidence without revealing any nervousness, whatsoever, even if it was there. For all of us it was a great relief after four days of rough sailing. We were very happy that it was time to see some action in the next 2–3 days.

On 27 December our ship was temporarily secured alongside a 2.0 m thick fast ice when we saw a group of Penguins surrounding the ship. It was an unforgettable scene that lingered in my mind for days. For the entire team it was a few minutes of mesmerizing experience where imaginations ran wild. Standing on the deck I was watching a vast sheet of ice with wind blowing gently in sub-zero temperatures pinching the exposed part of my face. As far as I could see it looked as if the sky is merging with the vast ice sheet and one cannot figure out where the horizon was! In Antarctica it was very difficult to make a guess about distances. In our normal life we develop a sense for distances by seeing buildings, trees, or any other objects. By their visual size, we estimate how far a landmark object could be. This is just not possible in Antarctica (Photos 2.6, and 2.7).

2.4.3 The Fateful Day of 29 December 1983

As soon as the ship was temporarily secured to the fast ice on 27 December, the Chetak helicopter was taken out. The first sorties by Chetak helicopter were made to view the proposed site of constructing the Base Camp. The Piston Bully was

Photo 2.7 A view of the base camp in Antarctica with the Chetak helicopter in the foreground

lowered with due precautions on the fast ice and 4 of us proceeded to inspect the proposed site that took a couple of hours to reach. We found it quite suitable as it was reasonably close to the shelf, where Finnpolaris would finally anchor to unload all the construction and other material amounting to some 1000 tons as well as unload all the oil barrels to run the diesel generators. The site was free of crevasses, smooth and above all had a very good view of Wohlthat Mountain Range located in the south. The Base Camp consisting of British Antarctic Survey (BAS) and medium Arctic tents was set up on the 28 December 1983. Chetak helicopters and Piston Bully were mostly used to transport material from the ship to set up the Base camp. A large amount of cargo was to be unloaded with the help of MI-8 helicopters. The stuff was packed in huge bundles and hooked to MI-8 Helicopter hoovering on the deck of the ship. If not properly accelerated, the underslung load could generate an undesirable swing hazardous for aviation. The fuselage of the MI-8 Helicopters was cut and a plexi-glass window was provided so that the flight engineer could view the under-slung load and advice the pilots accordingly (Photo 2.8).

It was 29 December 1983, I stood on the deck of the ship on a cold windy day when the MI 8 helicopter took off from the deck carrying an under slung load. To my horror I saw the helicopter struggling to gain height and while doing so the under slung load got entangled to the railing of the ship dragging the helicopter down and hard. In the process the rotor blades of the chopper hit the crane sending down the helicopter along with 5 occupants crashing into the icy cold waters of Antarctica. This happened so fast that it took some time to react as such an eventuality was not factored into our plans. My instant reaction was to set aside my feelings and without panicking a rescue mission was launched using a Chetak helicopter and a rescue boat. Standing on the deck and holding the railings tightly I was shouting instructions. To my dismay I saw the helicopter slowly drifting away from ship and started to sink. It was a stroke of luck that the helicopter overturned allowing the five occupants, Wing Cdrs. R. S. Tandon and B. D. Madhok, Flt. Lt. N. M. Rai, Sergeants S. B. Gupta,

Photo 2.8 MI-8 helicopter with the underslung load being carried to the camp site

and A. B. Jadhav to break open the plexiglass window in its fuselage and emerge on top of the sinking chopper. By that time the Chetak helicopter reached them ½ km away from the ship and hovered above them. But the worst was yet to come. When the first person was being hoisted to the helicopter, another crew member held on to his legs and gave a go-ahead signal, as staying alive in cold water was becoming difficult. As the helicopter lifted and started moving towards the ship, the person holding the legs lost his grip due to gusty winds and extreme cold and fell into the icy water. Just short of reaching the ship, the other person also lost the grip and fell. The exposure to icy water left them bereft of any strength. I was watching all this from a distance and my heart skipped a beat. But fortunately, the rescue boats that were pressed into service, reached the sinking helicopter and picked up the three almost drowned occupants of the helicopter. On way back to ship, the rescue boat also picked up the other two who had fallen halfway to the ship. The whole operation was indeed miraculous (Photo 2.9).

The doctors on the ship had been alerted about the accident and they were ready with emergency treatment. They were waiting anxiously to see the condition of those who had been in the helicopter and were worried about the likely effects of hypothermia on them. As soon as all the five arrived on board, they were rushed to the special hospital room prepared by the doctors. Looking at them I was in disbelief as barely an hour ago, they were bubbling with enthusiasm and energy and now they were in unconscious state. Hypothermia appeared to have set in due to the long exposure to sub-zero cold water. All those present on the ship were shell-shocked. I tried to console them. It was crucial at that moment to be cool and composed. It was necessary to inform Dr. Qasim of this happening. So I went to the Bridge of the Ship from where calls were made, and contacted Dr. Qasim. Several of my teammates and crew of the ship as well at the Captain of the ship were present when the call was made. Dr. Qasim was shocked to learn what had happened. He said that the Prime Minister needs to be informed, as she was keenly monitoring the progress

Photo 2.9 Chetak helicopter rescuing the occupants of the crashed MI-8 helicopter

being made in Antarctic-mission. Within a few minutes of my calling Dr. Qasim, I received a call from Mrs. Indira Gandhi and she enquired:

> How is everyone? Will they survive?

I explained the incident to her and said that I was hopeful of saving their lives. She heard patiently and then asked a direct question,

> Harsh, can you still do it?
>
> And my instant reply was,
>
> If I do not do it, I do not come back,
>
> There was a tense pause and then she said,
>
> Go ahead and do it.

What followed is a history. Everyone on the bridge heard my conversation with the Prime Minister of India. Captain Kulju who was pretty shaken by the accident and thought that I would call off the expedition as five of our personnel were in critical state asked for my instructions. I felt an adrenaline rush after the call that calmed me down and boosted my confidence to carry on with the mission with renewed vigor. All the team members heard my commitment to the Prime Minister, not to return till the task assigned was completed, no matter what. Seeing my steely resolve to complete the mission at any cost the hardened sailor gave his full support and was ready to go extra mile and provide whatever help was needed in those extraordinary circumstances. Immediately, a meeting was held with the team members who were present on the ship and decisions were taken on how to proceed with our construction work with reduced staff. Commodore Chandana advised me to meet other members of the team who were staying in the Base Camp and inform them about the accident. I made a visit to emergency room where doctors were treating the accident survivors before leaving for the Base Camp. Doctors told me that two of them were in critical condition at that moment but they were confident that their lives would be saved with medical equipment and medicines available to them in the temporary hospital that they had setup on the ship. I reflected on my decision to provide an exclusive room to them that came in handy to save the lives of the team members at a short notice. I was greatly relieved and thanked the doctors before I getting into the helicopter to reach the Base Camp.

Cdr. Chandana flew me to the Base Camp where work was in full swing, oblivious to the incident that happened a short while ago. Keeping a very brave face during the meeting that was arranged in the Mess Tent I narrated the incident and told them about the rescue operation and informed their status under the guidance of doctors on the ship. I also spoke about my conversation with Dr. Qasim and Prime Minister Indira Gandhi and my vow to her that I will complete the task or I shall not return. The initial expression of disbelief on the faces of team members turned into a sense of determination and they voiced in unison:

> We shall do it.

It was like a war cry. My respect and pride for the brave expedition members grew many folds and even today I have goosebumps when I recollect that meeting.

A message was received from the Air Force office, asking for setting up an enquiry about the MI-8 helicopter crash. A request was sent to the PMO's office to differ the enquiry as it would have taken precious time required for setting up the Base Station. It was a relief that the enquiry was differed.

2.4.4 Dakshin Gangotri Construction

It was necessary to develop a plan of action under the changed circumstances. On 30 December, a visit to the injured colleagues gave a lot of satisfaction, as they were responding well to the medical treatment. However, the condition of two of them remained critical. Now that we had a helicopter and five persons short a revised strategy was adopted. The under slung operations were discontinued to transport construction material to the base camp. Instead, boxes were unpacked and the material was carried in the fuselage of the remaining one MI 8 and two Chetak helicopters. This way the number of sorties was increased putting more work on Major Nair who took it as a challenge in the changed circumstances. He told me,

> The commitment made to the Hon'ble Prime Minister shall be honored no matter what ever happens. We shall work harder to achieve the target of setting up of Dakshin Gangotri in time and populate it with winter team members before the expedition leave Antarctica.

We planned to build the structure on a 620 m^2 raft foundation at a depth of 1.5 m so that there is even distribution of load to avoid any differential settling of the individual blocks of the building constructed on shelf ice. The building consisted of two double storey blocks of prefabricated timber structure, specially designed for Antarctic conditions that can stand winds of up to 100+ knots. The two blocks with sloping roof, forming loft for storage, linked by a narrow single storey corridor were planned. The prefabricated load bearing panels were basically made of wood. On top of the insulation material, metallic sheets were to be bolted. The radio room had acoustic tiles. The generator and the boiler room were provided with asbestos lining. The bathroom and kitchen had melamine—faced plywood. Block A was being built to house essential services such as various workshops on the ground floor and storage space on the first floor. Block B had laboratories, kitchen, lounge and living room on the ground floor and two-men bunkrooms, radio room and toilets for the 12 wintering team members, on the first floor. Three 62.5 KVA generators were to be provided for power requirements for the wintering team.

The construction began at a hectic pace and against all odds. It was −14 °C when we started digging trenches for raft foundation. It took 10 days in January 1984 to dig and clean up trenches for foundation. It was heartening to see that every individual was highly motivated and passionate about completing the assigned task. There was no difference between officers and men and it was a rare bonhomie one sees when there was a war cry for the country. The team of Major Pavan Nair was extremely dedicated and the 16 scientists equally participated in the construction work, whenever they had free time. Living in tents at the construction site was a very enriching experience. As per the Antarctic convention, at least one member remained awake while others slept. The communication with the outside world was only through satellite phones with limited time allotment. Every 3–4 days' team members used to change from Base camp to ship for a hot bath and a good sleep. There were also cases of cold injuries. I would then send the injured person to work from the ship and get a replacement. After a few days, no one reported of cold injuries. The generators were switched off at 10 pm, so that work could be halted, and one could eat and rest. The enthusiasm to complete the task in time was so high, that if the generators were not switched off, colleagues would not stop working. There were several days of white outs that hampered the construction work and slowed down our activities. By 17 January two blocks of the Wintering station were completed till the roof level (Photo 2.10).

By 21 January the Ship finally reached the shelf and shifting the material from the Ship to the Station became easier with the help of sledges and Piston Bully (Photo 2.11).

On 26 January 1984 I hoisted the Indian Flag with great pride and invited Scientists from Russian and East German Antarctic stations located nearby. The ceremony was kept very brief and the work was soon resumed. It was a proud moment for me when the Soviet Station Leader commented that I am leading a team of very brave people who are near to completing an impossible task of establishing the Indian station in one Antarctic summer and populating it. He said he is not aware of any such achievement before by any other country (Photo 2.12).

On 4 February three generators were switched on and tested and to our great satisfaction all of them functioned normally. The weather from 15 to 23 February was pretty bad with blizzards and whiteouts. But still we could make all the machinery,

Photo 2.10 Laying of the 620 m^2 of raft foundation for the base station

Photo 2.11 An arial view of the construction

Photo 2.12 Hoisting the National Flag of India in Antarctica on 26 January 1984

generators and heating systems functional. The Captain of the ship was pressing us to hurry up and windup operations as the fast ice had started forming and it was required that we move out to avoid getting stuck up there. He had given 1 March as the deadline for our departure to avoid any chance of getting the ship ice locked.

While the work to complete the primary objective of establishing Dakshin Gangotri was going on, a few of our scientists could find time to conduct front line research in Antarctica. As a part of a scientific program, a camp was established at Schirmarcher Hills 60 km away for geological sampling. A team of 16 sixteen scientists worked on varied fields like Meteorology, Communication, Geology, Magnetic surveys, Biology of krill and other Zooplankton, Sea Bird observations, Biological and microbiological studies, and hydro-chemical studies. A geological map on 1:25,000 scale and total intensity magnetic mapping of Schirmarcher Hills was prepared. A visit to Wohlthat Mountains was also made for future work. By 24 February 1984 Dakshin Gangotri became fully functional with all laboratories and facilities operating satisfactorily. But still the doubt lingered in my mind whether the structure would survive the harsh winter of Antarctica and whether it is safe to leave the wintering team of 12 scientists there, to be on their own. Dr. Qasim had left the decision to me and it was my call for which I will have to own up the consequences, either way. It was indeed a very heavy burden on my shoulders and I had a little time to take a decision.

It so happened that during 25–28 February we were about 40 people in the Dakshin Gangotri station when a blizzard hit the shelf. The wind speed exceeded earlier recorded highest speeds of 70 knots for this part of Antarctica, almost touching 80 knots. Those four days I kept my fingers crossed and took a sigh of relief when I inspected the station on 28 February. The Indian station withstood the blizzard and was standing in its full glory covered with ice all around. There were very minor problems that the maintenance engineers fixed immediately. Lt. Col. Sharma, the Station Commander and the Leader of the Wintering Team assured me that everything is going to be fine. Seeing the spirit of the wintering team I felt reassured and it made my life easier to take the final decision. The Captain once again pressed the need to set sail on 1st March. I thanked my stars that 1984 was a leap year and had 29 days in February. It was decided that Indian Team would leave Antarctica on 1 March 1984 leaving behind 12 members of the Indian Wintering Team. All requirements for the entire year were shifted from the Finnpolaris to Dakshin Gangotri and stored in their appropriated places. I doubt that anybody slept on 29 February's night. It was a night of anticipation, excitement, fear and joy. For me, I was turning in my bed every minute imagining, if something happened to the wintering team or facilities do not function as per requirement, if the roof collapse or a medical emergency occur, whether Dr. Ganguli will be able to handle, if there is fire or the station gets blown off in a stronger blizzard. All these thoughts continued to haunt me. The greatest satisfaction while leaving Antarctica was that all my team members who were involved in the accident were with me and recovering well.

Before departure the entire Wintering Team was brought to Finnpolaris for detailed discussions. It was clear that once the Finnpolaris sails off the Wintering Team had to fend for herself. The recent blizzard had provided me the strength to take the decision

Photo 2.13 Me in the cabin of Finnpolaris

to leave the Wintering Team at Antarctica. There was a grand beer party on the deck of the ship in which our friends, who were involved with the MI-8 helicopter accident also joined. That delighted all of us. Finally, we embraced and patted the members of the Wintering Team and bid them goodbye. Finally, we departed at 03.20 pm on 1 March 1984, leaving behind 12 brave Indians to face the first Indian wintering in Antarctica and fend for themselves for nearly 10 lonely months before the arrival of the next expedition in December 1984. We were all very proud of setting up the first wintering station in one Antarctic Summer campaign and creating a world record (Photo 2.13).

2.4.5 Journey Back Home

During our return passage, I kept talking to Lt. Col. Sharma several time a day. There were minor problems that the wintering team was able to take care.

I discovered that a few of our colleagues had suffered from cold injuries while construction of Dakshin Gangotri was going on. As a matter of fact, when the construction was in progress, and anyone reported of a cold injury, I would send him to work on the ship and get a replacement. I asked them, as to why did not they tell me. The answer was, "Sir, if we told you, you would send us to the ship. We wanted to continue to work to achieve the goal of setting up the wintering station in time." This brought tears in my eyes knowing their commitment to the mission, and I just kept hugging them for a long time. Journey to Mauritius was to take two weeks and we often had rough weather at sea but the pleasure of accomplishment and the feeling of going back home safely overwhelmed us and we did not bother much about the weather. On 8 March at about 6 pm, Lt. Col. Sharma called me to tell that they faced another blizzard where the wind speeds reached 80 knots and two anchor straps had broken. However, they were able to fix the straps after the blizzard was over. I again talked to him on 9 March. Things had settled down at Dakshin Gangotri. Days passed enduring rough sea, writing reports, talking to our family members and having rounds of 'Bridge' card game. Finally, we arrived in Mauritius at 02.00 pm on 18 March 1984.

2.4.6 Stop at Mauritius During Return Journey

I was getting regular feed from the wintering team at Dakshin Gangotri that things were settling down slowly as the 6 months dark phase began. A routine including scientific observations, feeding the melt tank, bringing the oil for daily use from the oil dump and above all adequate exercising in the gym was set in. Entrance to Dakshin Gangotri needed daily clearing of the drifted snow. India's success in setting up of Dakshin Gangotri at Antarctica in one Antarctic summer was very widely covered by the Mauritius news media. The entire team spent time visiting several beautiful places in Mauritius. Many were keen just to sit in a restaurant and eat something different after eating the frozen food on the Finnpolaris ship for the past several months. India's High Commissioner to Mauritius, Dr. Prem Singh had planned a very grand program for us. Prof Bhujdhar of Mauritius University had organized a talk by me to be given at the University on 19 March. I took help of Pietro, Chandna, and Puri to put together an illustrative talk on VCR. 19 March 1984 shall remain a very memorable day for us as we got ready to meet Dr. Prem Singh who was to take us to His Excellency Sir Seewoosagur Ramgoolam, the then Governor General of Mauritius. A very lively person at the age of 84 years lived in a palace with gardens sprawling over acres. He welcomed us very warmly. A very pleasant conversation followed. The Governor General was a very well-read person. He was keen to learn the secret behind our success. I told him that I believed the secret indeed was determination and total commitment and co-operation of the entire Indian Expedition Team very well supported by the Captain, and crew of Finnpolaris. It was a pure team-work and dedication of every individual, performing much beyond the tasks assigned to them.

We then met the Prime Minister of Mauritius, His Excellency Anerood Jugnauth. I presented him a piece of granitiferrous rock from Antarctica. Our talks spread over a variety of topics including the potential of Geothermal Energy in Mauritius. I was very impressed with the homework that his assistants must have done in digging out my book on "Geothermal Resources: An Energy Alternative" published by the Elsevier Scientific Publishing Company in 1980. The Prime Minister was very keen on developing and harnessing the geothermal energy potential of Mauritius. Our meeting lasted for almost an hour, far exceeding the original schedule of 30 min. After the meeting we had a small lunch at a Chinese Restaurant, and I found that I was getting late for the scheduled talk at the University of Mauritius. The talk was presided over by the Honorable Minister for Agriculture Mr. N. Deen Pal Singh. The talk went very well. It was followed by playing our VCR recording. Several

Photo 2.14 Meeting with the Prime Minister of Mauritius on 19th March 1984

interesting questions were asked. I found a bonding with the young students and faculty as well as very young ministers of Mauritius that I had been meeting the whole day. The Prime Minister was kind enough to sit for the entire 90 min of the talk followed by questions and answers. He complimented us and we waited for his departure before leaving the university (Photo 2.14).

The next day India House organized an evening. Dr. Prem Singh had invited several dignitaries. The Prime Minister of Mauritius had adjourned the Assembly so that the entire Cabinet Ministers could participate in the reception hosted by the India High Commission at the India House. There was a long photo-shoot with His Excellencies the Governor General, the Prime Minister, and several high-status socialites who were present. I introduced Major Pavan Nair, the main architect behind the success of constructing Dakshin Gangotri and other scientists who had done a remarkable work. I requested the Prime Minister to visit the Finnpolaris next day and Cdr. S. K. Chandna picked him up at 9.30. It was a great moment to welcome the Prime Minister of Mauritius accompanied by Finance Minister, Energy Minister, and Dr. Prem Singh. The Captain of Finnpolaris was very happy with the visit and gave a guided tour of the ship. The visit lasted for about 30 min. The entire Port Louis media was there, with live coverage and photo shoot. After a very satisfying visit, the Prime Minister and his colleagues were flown back. Later, we had a television interview for which Mr. Sahay accompanied me. A very pleasant stay in Mauritius came to an end on 22 March 1984 as we sailed off at 02.30 pm in the afternoon. From Mauritius it took us a week to reach Goa on 29 March 1984. Dr. Qasim was eager to meet us and know our welfare and without waiting for ship to arrive at the port of Goa.

2.4.7 Back in India

We were to reach the Goa port on 29 March 1984. Dr. Qasim wanted to meet us all on board Finnpolaris. Chandna brought him on Chetak helicopter. When ship reached the port we were told that in spite of a strike by the port workers they were present to welcome us while the naval band played on the welcome tunes. It was very heart-warming to see all the expedition members meeting their families. There was a grand reception in the evening, press conferences and a very wide coverage in the newspapers and television channels. Messages of appreciation poured from all over the world. Friends associated with Antarctic programs at UK and Soviet Union, whom I had been contacting and who had given me some very useful and practical tips were delighted with our success. I was informed that a presentation before the Prime Minister Indira Gandhi has been scheduled on the 6 June 1984. I took a sigh of relief as I had time for preparing myself and informed the expedition team. On 6 June 1984 the 3rd Antarctic team was present at Vigyan Bhavan at 6 pm to meet the Prime Minister. This was the first time we were meeting after our return form Antarctica. I prepared a 10-min presentation for the Prime Minister. The same was shared with Dr. Qasim and he concurred. The schedule for the Prime Minister's visit included a 10-min presentation, followed by some observations and questions, and interaction with the expedition members. This was to be followed with a cup of tea. The Prime minister arrived at the scheduled time. She had been very pleased with the success of the expedition in setting up of Dakshin Gangotri in one Antarctic summer and the wintering observations having been started by India. My presentation went quite well. Then the Prime Minister mingled with the expedition members sharing their experiences. The entire cabinet colleagues were with her. She kept interacting with the expedition members while tea and sandwiches were offered. She left almost an hour later than the original planned schedule. We were all very happy to see the Prime Minister taking so much time out of her very busy schedule and encourage us. Later, I came to know that when she was meeting us on 6 June 1984, Operation Blue Star was going on in Punjab. I realized that maybe she stayed longer with us to keep her mind diverted from that unfortunate incident happening at the Golden Temple in Amritsar (Photo 2.15).

Photo 2.15 Presentation of the 3rd Indian Scientific Expedition to Antarctica team members to Hon'ble Prime Minister of India Smt. Indira Gandhi on 6 June 1984

Photo 2.16 Presenting the first Indian wintering team to Mr. Rajiv Gandhi, the Prime Minister of India on 2nd July 1985

It was a very sad day for us when Prime Minister Mrs. Indira Gandhi was assassinated on 31 October 1984 at her residence in Safdarjung Road in New Delhi, by her bodyguards. Over the past 18 months or so, there had been close interactions with her. She had encouraged, supported, and made India's dream of contributing to Antarctic Science come true. The months to follow were full of anxiety. The wellbeing of the 1st Indian Wintering Team at Dakshin Gangotri was of utmost concern to me. After all, it was a decision taken by me to leave the Wintering Team in Antarctica. Quite often, it was depressing when Lt. Col. Sharma informed about difficult times that the team was facing. Gradually, Antarctic Winter months were getting over and Sun had started showing up. The 4th Indian Scientific Expedition to Antarctica was led by a good friend and colleague, Prof. B. B. Bhattacharya, from Indian School of Mines, Dhanbad, India, where I had studied for my master's degree in applied Geophysics. The 4th Expedition landed in Antarctica on 24 December 1984 and he informed me that all the 12 members of the team are healthy and the Dakshin Gangotri station has survived the Antarctic winter of 1984. That day I took a sigh of relief. The wintering team returned along the 4th Expedition summer team on 25 March 1985. A meeting of the wintering team was held with Prime Minister Rajiv Gandhi on 2 July 1985 where he suddenly asked me the reason of the helicopter crash. I did not have an answer and he just left with a mysterious expression on his face as if he knew the reason, being a pilot himself (Photo 2.16).

2.4.8 *Afterthoughts on the Success of the Mission*

In the hindsight, I factored some of the points that contributed to the success story of the establishment of our mission. Most important of them are:

1. Excellent interactions in the preparatory phase with the experts from U K and Soviet Union who had rich Antarctic experience and their advice, that proved very helpful.
2. Loading and unloading sequence of the equipment in the hold of ship that allowed us to save time in Antarctica; the equipment needed first was loaded in the last and vice versa. This was crucial as we had only a 35-day window to work.
3. The presence of the plexiglass window in the bottom of the fuselage of the MI 8 helicopter that saved the life of 5 occupants as they could break it open to come out of the chopper before being rescued.
4. Allotting a room exclusively for medical emergency with all medical facilities on board Finnpolaris was an extremely crucial decision that saved the lives of the 5 expedition members.
5. After the chopper crash, the commitment given to the Prime Minister of India and her extending full support came as a shot in the arm and led to a 'war cry' before any attack. This added a great vigor amongst the team members to complete the task and then only return back.
6. After the MI 8 Helicopter crash, there was a demand from the Indian Air Force to initiate an enquiry into the causes of the crash. Fortunately, on my request to the Prime Minister's Office, the same was differed till our return to India. It was a very positive, re-assuring and time saving decision. We were all glued together to do the job.
7. Despite the completion of the Dakshin Gangotri by the last week of February 1984, it was the blizzard that provided an excellent opportunity to get Dakshin Gangotri tested against the worst case scenario. It provided me the courage to take the decision to leave the wintering team for the Antarctic winter of 1984.
8. On the return voyage, I noticed that a few of my colleagues, who had the responsibility of constructing Dakshin Gangotri, had cold injuries. When I enquired why did they not report about it, the answer was that had they informed me, I would have sent them back to the ship; that they did not want and somehow continued the work with injuries. Such commitments and dedication of these brave young men of Indian Army was very crucial to complete the task in time.
9. The hard work of 81 members of the 3rd Scientific Expedition, the support of the Finnpolaris Captain and Crew and above all the backing and support of Dr. S. Z. Qasim and the Prime Minister of India made this almost impossible mission come true in just 35 working days available in the hostile environment of Antarctica.

Dakshin Gangotri propelled India's stature in polar research globally. The Indian dream of setting up of the first permanent station at Antarctica led to substantial increase in India's contribution to the science of Antarctica, addressing the problem of ozone hole, global efforts to restrict global warming and above all establishment of National Centre of Antarctic and Oceanic Research (NCAOR) at Goa by the Department of Ocean Development (DOD). In 2006, DOD was upgraded to a full Ministry of Earth Sciences (MoES). NCAOR has done India proud by pioneering work in Antarctica, Arctic and now in Himalayan glaciers as well. It was re-named as National Centre for Polar and Oceanic Research (NCPOR) in 2008, as work

Photo 2.17 Dakshin Gangotri postal stamp

Photo 2.18 More than half submerged Dakshin Gangotri in ice and snow, in 1988

had been also undertaken in the Arctic and the Himalayan Glaciers. India and the Indian Scientific community are reaping the rich benefits after signing the Antarctic Treaty in 1983 and very soon received the consultative status in the same year. Later, India joined the Protocol on Environmental Protection to the Antarctic Treaty (the Environmental Protocol), which entered into force for India on 14 January 1988. Over the years, India is globally recognized for her contribution to Antarctic research. The part played by India in the International Polar Year during 2007 was much appreciated. Towards the end of twentieth century, Ozone Hole was discovered over Antarctica, and it was established that the emission of halogen source gasses is primarily responsible for the creation of Ozone Hole. Persistent efforts to decrease their emission over the past two decades have made a difference. In 2019, contraction of the Ozone Hole was observed (Photos 2.17, and 2.18).

> The MI-8 helicopter accident on the 29 December, 1983 just on the third day of arrival in Antarctica appeared to totally disrupt our Antarctic Mission. On the contrary, it glued us together and our commitment not to return to India without completing the task, turned out to be a boon in disguise.

2.5 CESS Campus in Kerala, India

After finishing all formalities of debriefing of the 3rd Indian Expedition team and making presentations before Prime Minister of India, the unfinished task of creating a permanent campus for CESS in Trivandrum, Kerala, became the top priority. Lot of time went into planning the campus in the 19 acres of land allotted by Kerala Government in Ulloor area on Akkulum Road that was an abandoned clay-mining site. Particularly, care was taken that the campus should merge with the undulating topography of the area rather than stand as an eyesore. Buildings were designed keeping in mind the natural gradients, slopes and excavated mining areas existed at that time. A 5-block structure was designed with the main block to host the administration, library, conference hall and an auditorium. Four interconnected blocks were to provide seating spaces for scientists. The laboratory block hosted main laboratories of all divisions with a centralized access to scientists of all disciplines. A separate building for canteen and guesthouse was to be built away from the main blocks. A total built up area of 50,000 sq ft. with a centrally air-condition auditorium of 30,000 sq ft. having a seating capacity of 220 persons was finalized at an estimated cost of Rs. 1.40 crores.

The construction started in full swing, and it was decided that all the bills needed to be submitted by the contractors on Monday to be settled latest by Friday of the same week. In case there were any items that are not clear, as per the negotiated rates in the construction contract, payment of only those items was to be withheld, and payment of all the rest of items should be made. This practice was meticulously followed, and the contractors were happy and satisfied. A couple of hours were spent every day at the construction site before I proceed with my daily routine of administration and pursuing scientific research. One day I came to know that Dr. S. Varadarajan, Director General of CSIR was in town. He was known to be a great infrastructure creator all his life. Not to miss this opportunity I persuaded him to visit our campus and sought his advice after showing him the entire lay-out. I was pleasantly surprised when he expressed his desire to take a tour of the ongoing construction and then offer suggestions. After our fruitful site visit, he enquired about the estimates of construction, when we were having tea in my office. When the final cost of construction was revealed to him, he had a surprise expression on his face, and he told me in a concerned voice that the final cost will escalate to over Rs. 2.0 crores. This, he said is from the experience of allocating funds for construction to the Directors of CSIR labs for similar works.

The construction at the campus was progressing fast and the office staff was positive and ecstatic about it. After all, it was a very long pending dream of CESS scientists to have their own laboratories, offices, and other facilities. For me it was a great satisfaction that all this could happen within 3 years of my joining as Director, despite taking out time to establish Dakshin Gangotri in the intervening period. Finally, the new campus was inaugurated by the Chief Minister of Kerala, Mr. C. Karunakaran in the presence of several dignitaries that included Mr. K. R. Narayanan, who later became the President of India. The dignitaries spent more than half an hour going through the entire constructed area, appreciating the construction with total regard to the nature and how a land ravaged by mining had been converted into a lush green campus for CESS. The Chief Minister paid glowing tributes while speaking in the new auditorium and said that he laid the foundation stone in 1984 and in 1986 he is inaugurating the fully constructed campus. He spoke very high of my leadership and commitment and thanked me for giving Kerala, a state-of-the-art world-class research center in Earth sciences. He made a special mention about the cost and said that it was very heartening for him to know that against an estimate of Rs. 1.40 crores the building was completed with a miniscule overshot to Rs. 1.42 crores. He said it speaks volumes about the honesty and integrity of the people involved in the project. The lessons learnt during the construction of CESS remained with me for the rest of my life and were very useful in several construction projects that I successfully completed, later in my life.

2.6 Cochin University of Science and Technology, Kerala

One success leads to another and the word-of-mouth travel faster at the speed of light. Before I could breathe a little, Mr. V. Ramachandran, Chief Secretary to the Government of Kerala shocked and surprised me with the news of my selection as Vice Chancellor of the prestigious Cochin University of Science and Technology (CUSAT), Kochi in recognition of my services to the state of Kerala. He said that the Governor, Mr. P. Ramachandran has insisted that I should take up this responsibility. Since, it came from such a high authority I had no excuse to turn down the offer. It was a *fate accompli.* But there was a twist to the story. Being on deputation from NGRI for 5 years there was a technical problem of joining CUSAT directly from CESS, without the approval from my parent department, the CSIR. Dr. A. P. Mitra, Director General of CSIR treated this as a special case, probably the first case of its kind at that time and paid handsome compliments on my selection as Vice Chancellor at a young age of 45 years and put the case before the Governing Council headed by the Prime Minister, Rajiv Gandhi. It was a moment of great pride for me when rules were set aside, and my case was approved on merit paving my extraordinary deputation to CUSAT as the Vice Chancellor.

But before I joined CUSAT, as promised, CESS was placed onto the international map when several international experts in Earth sciences were invited for a seminar. Scientists who delivered key-note addresses and interacted with staff included Claude Froidevaux from France, Seweryn Duda from Germany, Eric Robert Engdahl from USA and S. Miyamura from Japan. We also had Dr. Hari Narain, Prof. Jagdeo Singh, Prof V. L. S. Bhimsankaran and Dr. B. P. Radahkrsihna from within India. In fact, Dr. Radhakrsihna was so happy after visiting CESS that he conducted the next Council Meeting of the Geological Society of India in the campus coinciding with a seminar on Environmental Geology. Stalwarts like Sunder Lal Bahuguna, who was spearheading the 'Chipko Movement' in Garhwal Himalaya against felling of trees spent a considerable time at CESS. Another longtime pending project on the 'Resource Atlas' of Kerala was also completed during this time.

After fulfilling my commitments at CESS I moved to CUSAT, a prestigious technical university that is primarily a post-Graduate university with under-graduate program in Ship Building Technology. As it was common in those days, universities were the hotbeds of students' union activities, often with affiliations to some political parties. Kerala was passing through a very interesting political phase in 1987 when I took office of the Vice Chancellor. Mr. E. K. Nayanar, having won 1987 elections became the Chief Minister and Mr. C. Karunakaran was the opposition leader. One day during a short visit to Kochi, Mr. Karunakaran conveyed to my office that he wants to meet me. Not knowing the purpose, I was anxious but when we met, he said that he was looking for an opportunity to meet me and express his gratitude for the contributions to the state of Kerala because both, the Dakshin Gangotri and CESS campus were established when he was the Chief Minister. Knowing the affiliations of the student's union to different political parties he assured me that he would advise the students not to create any trouble for me and rather make use of my presence in more academic way. I was moved by this wonderful gesture and thanked him profusely that made my tenure very smooth without any major incidents.

There was a chance meeting with Prof. Yash Pal, Chairman of the University Grants Commission (UGC) who was looking for a venue to host the 77th Session of the Indian Science Congress in 1990. This is one of the most prestigious and the biggest science event in India held every year and is inaugurated by the Prime Minister of the country. He said it would be great if CUSAT host the event under my leadership and bring further glory to Kerala state. Before committing I had a meeting with students who overwhelming supported the idea and the Education Minister Mr. K. Chandrasekharan promised to extend all help needed to host this mega event and said that he would talk to the Chief Minister. I was surprised when the Chief Minister, Mr. E. K. Nayanar made a sudden visit to Kochi and asked me to meet him at a short notice. The meeting went on well and the Chief Minister stressed that it was ironic that Kerala being the most literate state in the country had never been a venue of Indian Science Congress ever in the last 76 years. He took a lot of interest in knowing the requirements of the Science Congress and asked me the expectations from his side. I explained all the details about requirement of the space for hosting 18 parallel sessions, accommodation for about 4000 participants, arrangement for food, Science Exhibition, and all related matters.

The Chief Minister listened to me patiently and enquired about the finances for such a big event. In those days the Department of Science and Technology, Government of India used to grant a seed money of Rs. 1.0 crore (10 million) for the Indian Science Congress and it was the responsibility of the host to generate the remaining expenditure through exhibition, registration fee from participants and other sponsorship. I informed him that Prof. Yash Pal had also promised partial financial support in addition to some funds from DRDO. It was sheer music to my ears when Mr. Nayanar said that it was the turn of Kerala to show its appreciation to what all I did for them till then. He promised that Kerala Government would meet whatever is the shortfall in expenditure after exhausting all the sources. With such an unequivocal support of the government I made a call to Prof. Yash Pal and informed him that CUSAT would be glad to host the 77th Indian Science Congress in 1990.

In all earnestness and under the overall supervision of Prof. K. G. Nair, a senior professor we began the preparations. I got a call from Dr. A. P. Mitra, Director General of CSIR, inviting me to deliver a talk at the 76th Session of the Science Congress to be held at Madurai in January 1989 so that I can have a ringside view of the event next year in our campus. Dr. Mitra was the General President for the 76th Science Congress. Dr. K. G. Nair accompanied me and made detailed notes, photographed the venue and sessions. In the Science Congress it was customary that the Chief Guest, other dignitaries, past presidents of ISCA and Sectional Heads of the different scientific sessions wear Convocation Gowns and walk to the dais to occupy their designated place during the inauguration. To give a southern touch and a more traditional look we decided to do away with the Convocation Gowns and instead put a shoulder shawl. The idea appealed to everyone. We decided another change that there will be no separate enclosure for VIPs and the same food will be served for everybody. In the hindsight, I feel that these two decisions had a great impact on students and faculty as I could see a sea change in their attitude and involvement in hosting the Science Congress. To generate additional resources, the industrialists and the business community were approached, but the response was lukewarm. It was realized that this was due to the location of CUSAT being away from the center of city and a guaranteed attendance of the participants and the public for 5 consequent days was not assured. After several brainstorming sessions, a decision was taken to shift the venue of the exhibition to some place closer to the city and increase the duration of exhibition to one month, which was again a first in the history of Indian Science Congress.

The D-Day arrived when the Prime Minister of India Mr. V. P. Singh arrived Kochi on 4th February1990. As per the protocol, the Prime Minister was received at the airport and brought directly to CUSAT where I along with Prof Yash Pal welcomed him. As per the convention, there was a group photograph of the ISCA Council with the Prime Minister. We felt very happy when Mr. V. P. Singh appreciated and commented on the idea of wearing a shawl instead of conventional gown. I made a brief welcome speech, which was followed by the speeches of the Chief Minister, the Education Minister and then the Presidential Address by Prof Yash Pal who spoke on the theme of the 77th Science Congress: "*Science in Society*".

The Prime Minister delivered his inaugural address that was welcomed with a thunderous applause. He spoke extempore and with great passion and made an appeal to the young Indians to work together towards nation building and progress of humanity. Everyone was mesmerized by the brilliance and heartwarming speech of the Prime Minister. After the inauguration, the Prime Minister left the car that had brought him from the airport and instead sat in the front seat of the security jeep to go around CUSAT Campus and visit a few departments, including the Departments of Electronics that had been very heavily supported by the Department of Electronics, Government of India. After a brief stay the Prime Minister left for the airport to be in Delhi for his other engagements.

The 77th Indian Science Congress became famous for a couple of new initiatives: Firstly, all dignitaries wore a traditional shawl instead of conventional gowns, and, secondly, there were no separate enclosure and food arrangements for VIPs as all were served the same food. The venue of Exhibition was shifted away from the venue of Science Congress and a nominal fee of Re. 1/- head was charged as entrance fee and the exhibition duration was extended for a period of one month. In an another first, Prof. Yash Pal asked me to chair the concluding session of the Science Congress, which is generally chaired by the General President himself. As the response at the exhibition was overwhelming, substantial funds were generated despite a nominal entrance fee of Re. 1/-. These funds were later used in awarding scholarships to the deserving students. CUSAT has created a history for Kerala by organizing the first ever Indian Science Congress in the state where I played a major role and bringing glory to the state.

2.7 Earthquake Forecasts

2.7.1 Earthquake Forecast in the Northeast India

The most common question that was posed to me by media after each of the damaging earthquake in India was, *why can't the scientist predict earthquakes?* Though, there had been several claims of successful earthquake predictions, globally, these were not found to be true on scientific scrutiny. The classical case is the prediction of the magnitude of 7.3 Haicheng earthquake of 4 February 1975 by the Chinese seismologists. The success of the prediction of this earthquake continues to be debated even today. Earthquakes are difficult to predict in short term windows and any meaningful prediction should have a definite time window, a magnitude, and the location within a reasonable acceptable error limit. There is a global effort to study temporal and spatial variation of earthquake occurrence and recognize patterns that precede major earthquakes. Detailed investigations carried out globally, and particularly in New Zealand and California, demonstrated that often precursory swarms followed by quiescence occur before major earthquakes. Several precursory studies are being carried out to monitor geophysical, geochemical, and animal behavior to predict earthquakes. But the results have not been encouraging. Since, there was some improvement in earthquake monitoring in the northeast India, we thought that it would be a worthwhile exercise to look at the data sets and make out some sense.

An earthquake of magnitude 5.4 had occurred on the 30 December 1984 in the Cachar region in the northeast India. It was found to be preceded by a well-defined precursory swarm and then a quiescence. This encouraged us to take up a very detailed investigation of the north-east India seismicity during 1986. The first step was to prepare a unified catalogue of north-east Indian earthquakes. We found that 10 earthquakes of magnitude $\geq$7½ had occurred in the northeast India region since 1897. A detailed analysis revealed that these earthquakes were preceded by recognizable earthquake swarms and quiescence pattern. An empirical relation was developed between mainshock magnitude, the magnitude of the largest earthquake in the swarm and the duration of the swarm. It was realized that the most important issue is to recognize an area where swarm and quiescence have already occurred without a mainshock and then use the empirical relations developed to forecast the location, magnitude, and the probable time of occurrence of the next possible earthquake. An area bound by 21°N–25 ½°N and 93°E–96°E was delineated where a swarm had occurred during 1964 with 4 earthquakes of 6 and larger magnitudes. This was followed by a quiescence at the time of the research being carried out in 1986. Based on these and other observations, we concluded that:

1. Moderate to great earthquakes in the northeast India region are preceded by well-defined earthquake swarms and quiescence periods.
2. Based on an earthquake swarm and quiescence period observed in the data from 1964 to 1986, the next site of a probable earthquake would between 21°N and 25½°N latitude and 93°E–96°E longitude. The magnitude of the earthquake would likely to be of M 8 ± ½ with a focal depth of 100 ± 40 km. This earthquake should occur any time starting 1986 and should it not occur till the end of 1990, this forecast could be considered as a false alarm.

This prediction came true and an earthquake fulfilling this forecast occurred on 6 August 1988, as seen the following Table 2.1.

The probability of an earthquake of M ≥ 7½ occurring in the area and the time interval of prediction was 0.048 which is very small. Hence, the occurrence of the forecasted earthquake can be considered significant. In the IUGG General Assembly held in August 1987 at Vancouver, Canada, I had presented our results during an invited talk. Dr. Carl Kisslinger was chairing the session and he commented that it would be the second successful medium term earthquake forecast if the earthquake occurs. The proceedings of the Vancouver session were published in the Tectonophysics next year in 1988 and by that time the earthquake we forecasted had occurred. A note that is produced here was added at the end of the paper mentioning the successful forecast of this earthquake (Fig. 2.1).

Note: The occurrence of an earthquake of magnitude 7½ with an epicenter at 25.116°N, 95.171°E (focal depth 115 km, origin time 00:36:26.9 GCT on 6th August 1988 (parameters from NEIS. USA) as forecasted a year ago means that our medium-term forecast was accurate.

Buoyed by the success of this forecast in the northeast India we turned our attention to the seismically active Koyna region in Maharashtra, where triggered earthquakes were occurring since 1962. Though, there may be several unknown patterns, yet to be discovered, a case study of a process called *nucleation* is mentioned here that holds a great promise in forecasting earthquakes. We used this concept to make few successful forecasts in the region. This process has a sound experimental basis on rock failure mechanism as shown in several rock fracture studies.

Table 2.1 Earthquake forecast for the northeast India region

Earthquake parameter	Forecasted Gupta and Singh, 1986	Occurrence NEIS
Epicenter	21°N–25 1/2°N 93°E–96°E	25.116°N 95.171°E
Magnitude (M)	8 ± ½	7½
Depth	100 ± 40 km	115 km
Time	Feb. 1986–Dec. 1990	August 6, 1988

Table provided the forecasted parameters and the earthquake occurrence details

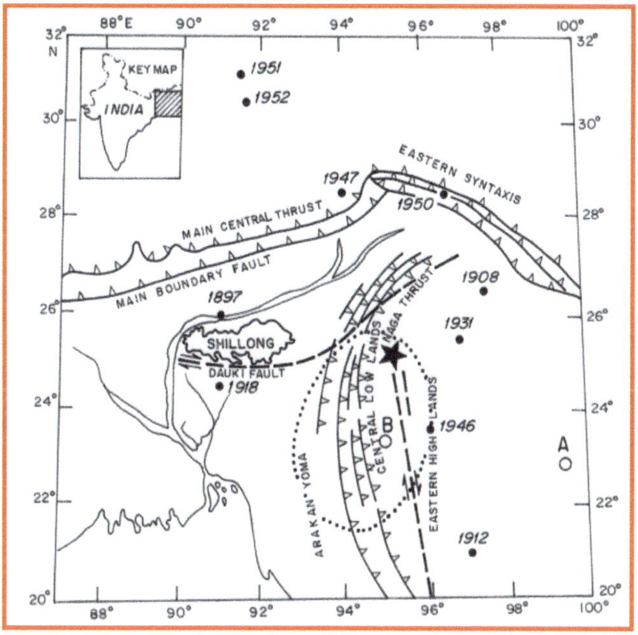

Fig. 2.1 Epicenters of M ≥ 7 and ½ earthquakes for the period 1897–1962 (black filled circles), forecasted area for the occurrence of M 8 ± 1/2 earthquake (dotted ellipse), and the epicenter of 6 August 1988 M 7½ earthquake (filled star) and two other aftershocks of M > 7 following the 6 August 1988 earthquake (open circles, A and B)

2.7.2 *Earthquake Forecasts in Koyna, India*

Koyna Dam, located near the west coast of India is a unique site of artificial water reservoir triggered earthquakes, where the earthquake occurrence is confined to a small area of 20 km × 30 km, at depths between 2 and 9 km. We postulated in 2001, in one of our publications, that short time earthquake forecast may be feasible at Koyna. In 2005 a dense network of seismic stations was deployed in the immediate vicinity of Koyna and the earthquake data from these stations were collected in real-time. From among all the observable earthquake precursors, study of foreshocks holds a great promise in making earthquake forecasts. M. Ohnaka, a Japanese seismologist in 1992 showed experimentally that the immediate foreshock activity constitutes a nucleation phase that leads to the occurrence of the mainshock in an earthquake sequence. To examine whether there are recognizable foreshocks preceding $M \geq 4$ earthquakes in the Koyna region, a detailed study was undertaken. It was observed that $M \sim 4$ earthquakes in the Koyna region are generally preceded by a well-defined foreshock sequence. Table 2.2 given below demonstrates the learning phase where four earthquakes of M 4.0–4.8 were preceded by a significant foreshock activity confined to a tight cluster within 10 km radius as shown in Fig. 2.2. This finding provided the much-needed confidence that a short-term forecast could indeed be possible for magnitude ~4 earthquakes if a nucleation phase is observed in real time.

Continuous observations of the earthquake occurrence and their location in real-time were undertaken. During the period 11–16 May 2006 a very well-defined nucleation was observed. Some 50 earthquakes of $M \geq 1.0$ with the largest event of M 2.7 confined to a focal depth of 2–8 km occurred clustered in a small area for 107 h. We realized that the region was experiencing a nucleation phase. This led to making the following forecast (Fig. 2.3):

> Based on the data available from seven seismic stations operating in the Koyna region, we have identified a nucleation which started on 12 May 2006. This may lead to the occurrence of an M~4 earthquake in the next 15 days. This shallow earthquake (focal depth less than 8 km) will occur within a radius of 10 km centered at 17.1°N, 73.8°E. Based on our previous experience of studying nucleation preceding earthquakes in the Koyna region, we expect this earthquake to occur over the next 15 days-time (till 31 May 2006), with a 50% probability.

This finding was communicated to the Editor of Current Science, President of the Geological Society of India and the Secretary of the Ministry of Earth Sciences, Government of India. Purposely, this information was not shared with the news media to avoid any panic. The forecast came true with the occurrence of an M 4.2 earthquake on the 21 May 2006 well within the forecasted parameters (Table 2.3).

After making a successful forecast of M ~ 4 earthquake in Koyna during May 2006, similar studies continued. In the years to follow the effort continued and a total of eight forecasts were successfully made. The magnitude of forecasted earthquakes ranged from M 4.0 to M 5. In a later study it was noted that during the period August 2005 through May 2010, a total of 29 earthquakes in the magnitude range of 3.5–5.1 occurred in the Koyna region, which could broadly be placed in three zones (Fig. 2.4). While Zone A has been the most active during this period accounting for 18 of these

Table 2.2 Identification of nucleation preceding M ~ 4 Earthquakes

Main earthquake	Duration of nucleation period (h)	No. of events before the main earthquake				
		M 1.0–1.9	M 2.0–2.9	M 3.0–3.9	Largest earthquake	50 h prior to beginning of nucleation period
30 August 2005 M 4.8	110	13	8	0	2.6	2
13 November 2005 M 4.0	160	18	8	1	3.0	3
26 December 2005 M 4.2	150	22	8	0	2.9	6
17 April 2006 M 4.7	380	40	10	0	2.7	2

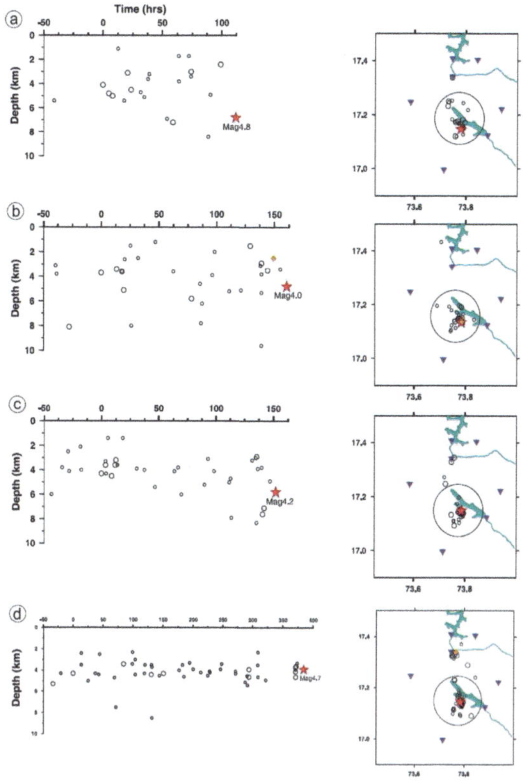

Fig. 2.2 Seismic activity for the period 50 h before the start of nucleation till the occurrence of $M \geq 4$ earthquakes in Koyna region during August 2005–April 2006. **a** 30 August 2005; **b** 13 November 2005; **c** 26 December 2005; and **d** 17 April 2006. Foreshocks clustering before the mainshocks in a region of 10 km radius (Right). Temporal distribution of events within the circle with the depth. o, O, and ★ are events in M 1.0–1.9, 2.0–2.9, and 3.0–3.9 ranges, and main earthquake, respectively (Left). Onset of identified nucleation period is given '0' time mark. Nucleation period lasted for 110, 160, 150, and 380 h for **a** 30th August, **b** 13th November, **c** 26th December 2005 and **d** 17th April 2006 earthquakes, respectively (Table above)

31 earthquakes, Zones B and C had 5 and 6 earthquakes in the 3.5–5.1 magnitude range. While earthquakes in Zone A are found to be mostly preceded by nucleation, the same is not the case with Zones B and C. It must also be mentioned that all M ~ 4 earthquakes were not forecasted and there was no false forecast. This amply supports the complexities in earthquake forecasting. Even in a small region, like Koyna, there are pockets of rock volumes where precursory nucleation occurs, but that is not true for the entire Koyna region.

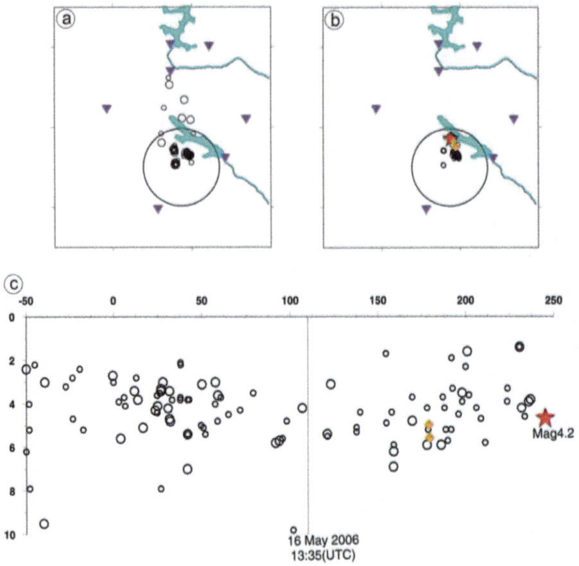

Fig. 2.3 Symbols as in Fig. 2.2. **a** Events that occurred during 107 h (11–16 May) preceding identification of nucleation. **c** Temporal-depth plot of the events within the 10 km radius circle. Nucleation was inferred to have started 107 h before, where 0 h is put on the time axis. **b** Seismic activity from 13:35 UTC on 16 May till the occurrence of the M 4.2 earthquake on 21 May 2006, on the right side of (**c**)

Table 2.3 Forecast of 21 May 2006 M 4.2 earthquake in Koyna on 16 May 2006

	Forecast parameter	Occurrence
Epicenter	Within 10 km radius of 17.1°N, 73.8°E	17.171°N, 73.777°E
Mag	~4	4.2
Time	Within 15 days of 16 May 2006, i.e. until 31 May 2006	20:29:01 UTC On 21 May 2006
Focal depth	Less than 8 km	4.7 km

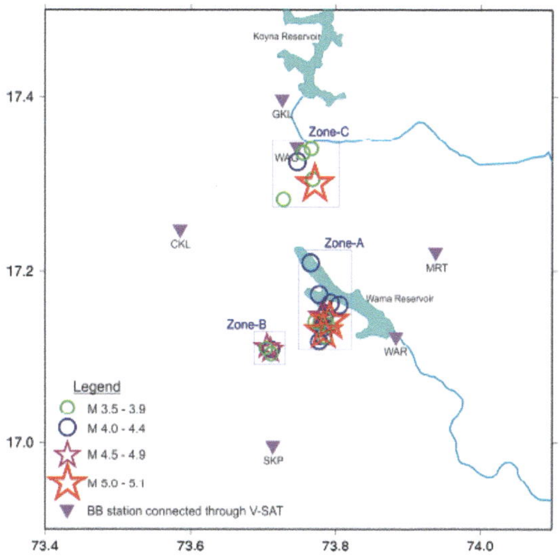

Fig. 2.4 Earthquakes of magnitude range 3.5–5.1 in the Koyna-Warna region during August 2005–May 2010. Zone 'A' indicates area in which all M 3.5 and above earthquakes preceded by nucleation, whereas in Zone 'B' and Zone 'C' nucleation was not observed

Chapter 3
Forays in International Arena

3.1 Leading Global Seismic Hazard Assessment Program

Chairing the Steering Committee of the Global Seismic Hazard Assessment Program during 1992–1999 was monumental, and very satisfying. To minimize the impact of natural disasters, the United Nations in the General Assembly held on 11 December 1987 declared 1990's to be the International Decade of Natural Disaster Reduction (IDNDR). The Global Seismic Hazard Program (G-SHAP), under the IDNDR umbrella, as a demonstration project, focused on seismic hazard assessment globally. The period of implementation of G-SHAP was from 1992 to 1999. It was supported globally by national scientific agencies and research institutions. Over 500 scientists worked from all over the world for the compilation of Global Seismic Hazard Map. The work was carried out region wise. I had the responsibility for India and adjoining regions. The G-SHAP implementation and activities were supervised by a Steering Committee constituted to overview the progress of the project and make suggestions for improvements. The Steering Committee had 20 global experts in seismology and earthquake engineering. I had the honor of Chairing it.

The work conducted under the G-SHAP umbrella focusing on technical planning and guidelines for assessment of earthquake hazard till 1993 was published in Annali Di Geofisica (Vol. XXXVI, No. 3–4, June–July 1993). A summary of the status of seismic hazard assessment globally was presented in this volume. This work provided the technical background for developing a unified approach for seismic hazard assessment and for preparing the global seismic hazard map. In addition to chairing the Steering Committee, I also had the responsibility for compiling the status of knowledge of the seismic hazard from Iran to Burma (Myanmar now) including Iran, Afghanistan, Pakistan, India, Nepal, Bangladesh, and Burma. All the available information, as of 1993, concerning existing earthquake catalogues, seismotectonics, strong-motion data, and seismic zoning maps was compiled and included in the publication ('Seismic hazard assessment in the Alpide belt from Iran to Burma', pp. 61–82). It was pointed out that the loss of human lives in these Asian countries

H. K. Gupta, *Harsh K. Gupta*, Springer Biographies,
https://doi.org/10.1007/978-3-031-63561-8_3

was much larger compared to the developed countries. The two major factors that contributed to this situation were very poor quality of constructions and high density of human population.

G-SHAP related work was carried out with several meetings, reviews, and demonstration projects. Finally, the work carried out under the G-SHAP umbrella during 1992–1999 was compiled and published as: 'The Global Seismic Hazard Assessment Program (1992–1999); Annali Di Geofisica (Vol. 42, No. 6, December 1999). In addition to Chairing the Steering Committee of G-SHAP, I was also involved in preparing 'A probabilistic seismic hazard map of India and adjoining regions' (pp. 1153–1164), and compilation of 'Global Seismic Hazard Program (GSHAP) in Continental Asia (pp. 1167–1190) (Figs. 3.1 and 3.2).

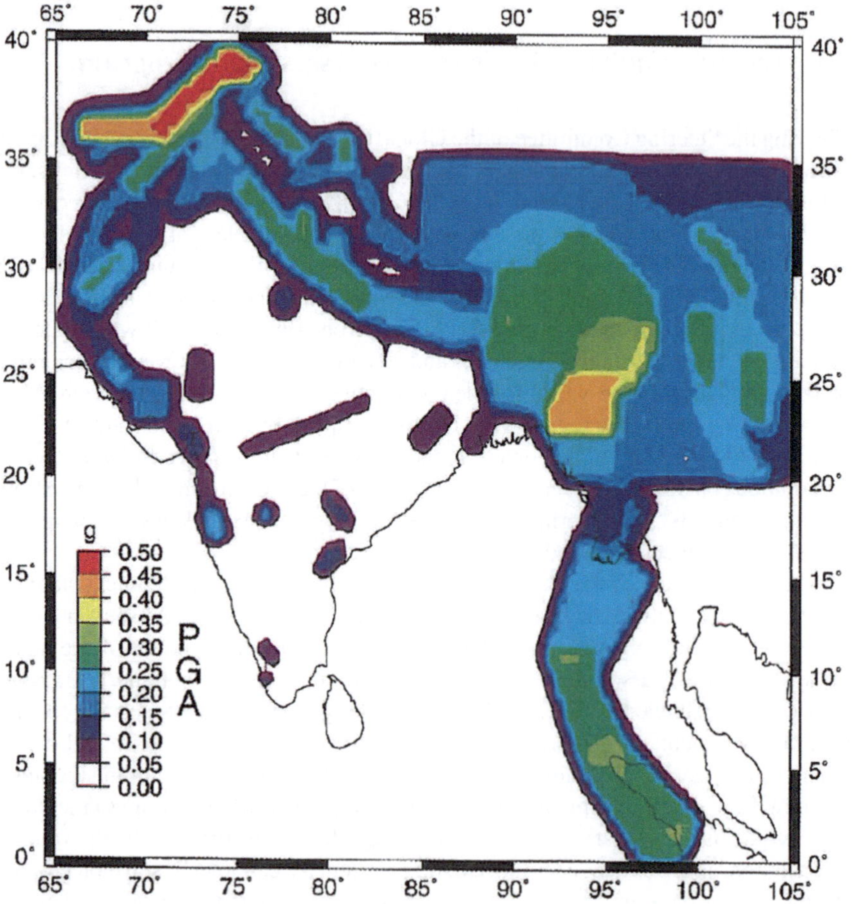

Fig. 3.1 Seismic hazard map of India and adjoining regions for 10% probability of exceedance in 50 years, contour interval 0.05 g

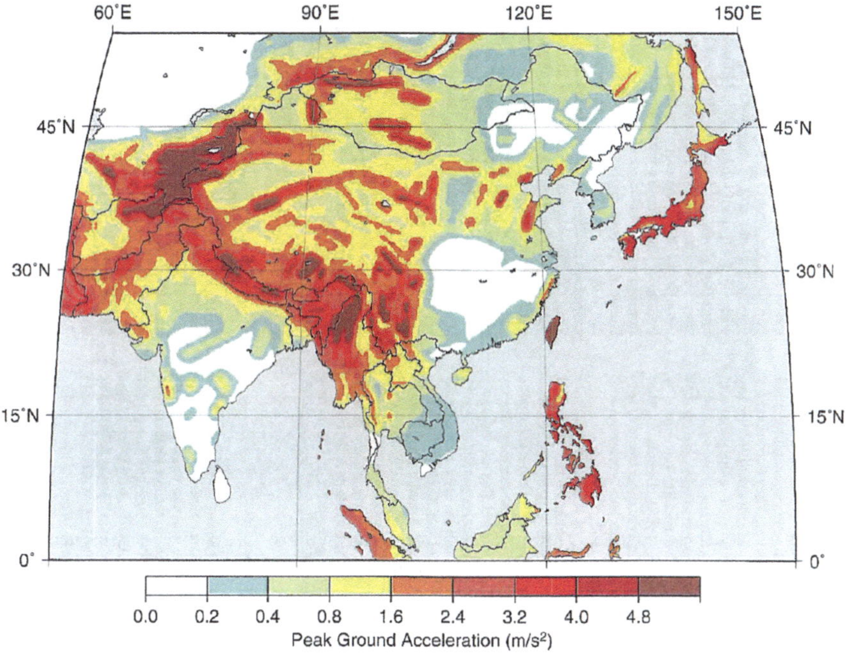

Fig. 3.2 The seismic hazard map of Asia depicting peak ground acceleration (PGA), given in units of m/s^2, with a 10% chance of exceedance in 50 years

It may be noted that G-SHAP improved global standards in seismic hazard assessment. It particularly promoted appropriate use of multi-disciplinary information, improved the earthquake data base, and helped in standardization of earthquake hazard knowledge. It was particularly helpful for the developing countries. It encouraged cross-border collaboration. The hard work of over 500 researchers for over 7 years ultimately led to a reliable global map of earthquake hazard. Among several global programs that I have had the privilege of being associated, the participation in preparing of the Global Seismic Hazard map has been a very satisfactory one.

While working as the Director of the National Geophysical Research Institute from 1992 to 2001, several important meetings were held, and I was lucky to be decorated with some rare honors. Here, I would just like to mention hosting the 1997 Rajiv Gandhi Memorial public lecture which was delivered by a famous French Scientist, Noble Laureate Jean-Marie Lehn. It was held at the Ravindra Bharti Auditorium, which is centrally located in Hyderabad, India. It was very well attended, and the lecture was presided by Mrs. Sonia Gandhi, Chairperson of the Rajiv Gandhi Foundation (Photo 3.1).

The other memorable event was my receiving the Federation of Indian Chambers of Commerce and Industry, New Delhi, award for the Physical Sciences for the year 1998–1999 by the then Prime Minister of India, Mr. Atal Bihari Vajpayee (Photo 3.2).

Photo 3.1 NGRI hosted the 1997 Rajiv Gandhi memorial public lecture delivered by French Nobel laureate Jean-Marie Lehn at Ravindra Bharti, Hyderabad. From left: Mrs. Sonia Gandhi, Jean–Marie Lehn, Harsh Gupta

Photo 3.2 Harsh Gupta receiving the Federation of Indian Chambers of Commerce and Industry, New Delhi, award for physical sciences,1998–99 from the Hon'ble Prime Minister, Mr. Atal Bihari Vajpayee

Photo 3.3 Harsh Gupta welcoming Dr. Murli Manohar Joshi, minister for science and technology at the NGRI during the CSIR director's conference in 2000

Another momentous event has been hosting the then, Union Minister for Science and Technology, Dr. Murli Manohar Joshi, at NGRI when he presided over the CSIR Director's Conferences in the year 2000 (Photo 3.3).

3.2 Formation of the Asian Seismological Commission

The Latur earthquake struck the Stable Continental Region (SCR) of Peninsular India on 30 September 1993, just after one year of my taking up the Directorship of NGRI at Hyderabad. It was a moderate earthquake of 6.3 magnitude but killed more than 10,000 people, mainly in Killari and the surrounding villages. A preliminary report on the Latur earthquake published in Science entitled 'The Deadly Latur Earthquake' (Science, Vol. 262. 10 December 1993, pp. 1666–67) drew worldwide attention. It was the deadliest Stable Continental Earthquake till 1993.

The 1990s was the period of change in India when Digital Broadband Seismographs networks replaced the old analog seismographs leading to availability of high-fidelity data that brought a quantum jump in the way the seismology research is done in the country, today. It is now at par with global Digital Broadband Networks that has become a reality in the last few decades by the efforts of international organizations like IRIS, GEOSCOPE, GEOFON, OREFUS, CDSN through installation of broadband 3-component digital systems. This effort was supported by advances in data transmission and dissemination coupled with INTERNET revolution. This has strongly facilitated in bringing together various scientific groups of diverse expertise in varied disciplines to address geodynamic problems of global interest.

The Latur earthquake generated a great interest amongst the global scientific community because the causative fault of the earthquake ruptured the surface, a very rare phenomenon in a Stable Continental Region (SCR) environment. The world-renowned seismologist made a beeline to the earthquake affected area to see this rare feature. Post Latur, NGRI led a major World Bank assisted project to set up a Broadband Seismic Network in the Indian shield region with the participation of research institutes, universities, and other organizations to develop human resource in seismology. NGRI established three Broadband Seismological Observatories at Cuddapah, Kothagudem and Dharwad. Initially, these observatories were connected to a Central Recording Station (CRS) at NGRI via VSAT link for online data transmission. Presently, the data from this network is transmitted online to National Centre for Seismology (NCS), New Delhi that was carved out of IMD and attached to Ministry of Earth Sciences.

On 22 May 1997 another moderate earthquake of Mw 5.8 struck Jabalpur in Central India and claimed a few tens of lives. The very next day NGRI established a local network of 5 seismic stations around the epicentral region to study the pattern of aftershocks. Later, data from NGRI and GSI networks lead to the first detailed study on the microzonation of Jabalpur city based on local site geology and data from local network. Two damaging earthquakes in a SCR environment in short span of 5 years attracted international attention and led to very prestigious theme based *Chapman Conference* on 'Stable Continental Earthquakes' hosted at NGRI during 25–29 January 1998. It attracted 90 researchers from 12 countries. Detailed discussions were held and the proceedings were summarized in Eos, Transactions, American Geophysical Union (Vol. 79, No. 27, July 7, 1998, pp. 319–321). It was realized that the Stable Continental Regions are more vulnerable to earthquakes than once thought.

On several occasions, both nationally and internationally it was pointed out that the Asia–Pacific region has suffered 70% of the human lives loss during earthquakes and the numbers have been increasing. Added to this are the financial losses that grew exponentially due to earthquake disaster. An opportunity arose to create an Asian platform for scientists to exchange ideas and experience during a seminar on East Asia seismological studies in Tokyo in late 1993. China, the other country where human losses during earthquakes had been very high was also looking forward to such a platform under the leadership of Prof. Zhu Chaunzhen, known for the first successful earthquake prediction in China in late 1960s. Delegates from 15 countries that included India, China, Iran, Japan, Korea, Malaysia, The Philippines, and Thailand attending the seminar suggested formation of an Asian body under the aegis of International Association of Seismology and Physics of the Earth's Interior (IASPEI), one of the seven associations of International Union of Geodesy and Geophysics (IUGG).

Everybody present in the seminar welcomed the idea and an ad hoc Working Group under Prof. Zhu was formed. It was agreed that he will make a presentation before the IASPEI Executive Committee (EC) meeting in 1994 chaired by President Prof. A. Nikolaev. After listening to the proposal, the EC agreed to the formation of the Asian Seismological Commission (ASC) as one of its Regional Commissions under the umbrella of the IASPEI and suggested to expand it to include Asia and Oceania regions that will cover Australia and New Zealand.

Following the support from IASPEI, I was invited to Guangzhou in 1995 for a preparatory meeting of the ASC where delegates from China, India, Iran, Japan, Korea, Philippines, and Vietnam participated. Dr. Eric Robert Engdahl, the Secretary General of IASPEI announced the support of IASPEI to ASC both academically and financially. Later next year, the IASPEI Regional Assembly in 1996 commemorated two decades of 1976 Tangshan earthquake of China and declared the formation of ASC. I was elected as the First President of ASC with Prof. Zhu Chaunzhen (China) as Secretary General and Profs. H. Hamada (Japan) and G. Gibson (Australia) as the two vice presidents of the founding ASC Bureau.

More than 380 delegates attended the IASPEI Regional Assembly that included participants from USA, Europe, and other western countries. Asia and Oceania region was represented by participants from 19 countries including Australia, Azerbaijan, Bangladesh, China, India, Iran, Israel, Japan, Kazakhstan, Korea, Kyrgyzstan, Nepal, New Zealand, Pakistan, Russia, Thailand, Philippines, Turkey, and Vietnam. Since, the ASC was first announced in 1996 IASPEI Assembly, it was considered as the 1st General Assembly of ASC. It was agreed that the General Assemblies of the ASC will be held once in two years in some Asian country.

After taking over as the First President of the ASC the next General Assembly was organized at NGRI, Hyderabad during 1–3 December 1998. The response was overwhelming as it attracted 189 abstracts and 200 delegates attended the Assembly with 70 participants from outside India representing 26 countries. A 15-Day International Training Course on "Seismic Monitoring, Data Analysis and Exchange" preceded the General Assembly during 15–30 November 1998. 25 participants from China, Iran, Oman, Uzbekistan, Kyrgyzstan, Georgia, Zambia, Yemen, and Indonesia were trained at NGRI.

The beginning of the 21st Century had a surprise for India as the Bhuj earthquake of magnitude 7.7 occurred in the morning of India's Republic Day on the 26 January 2001. It gained the reputation of being the deadliest Stable Continental earthquake having claimed over 20,000 human lives, surpassing the lives lost in the 1993 Latur earthquake. A report in Science summarized the lives lost and damage caused by this earthquake ("The Deadliest Intraplate Earthquake", Science, 16 March 2001, V 291, pp. 2101–2102). It may be mentioned that Dr. R. K. Chadha, a Scientist from NGRI, was elected as the Secretary General of the Asian Seismological Commission in 2004. He had three consecutive terms and provided outstanding service to ASC till 2017.

The Asian Seismological Commission has been serving the Asia–Pacific region providing a platform for discussions on earthquake related issues and finding a way forward to reduce earthquake hazard. In 2021 ASC completed 25 years. It was commemorated by bringing out a dedicated issue of the Journal Geological Society of India, entitled "25 Years of Asian Seismological Commission" (J-GSI, Vol. 97, No. 12, December 2021, pp. 1475–1602). It has a total of 19 articles authored by global leaders highlighting the work carried out during the 25 years of ASC in different aspects of earthquake studies. In conclusion it may be said that setting up of the Asian Symbolical Commission has been very successful and it has served the humanity (Fig. 3.3; Photo 3.4).

Fig. 3.3 Logo of ASC, designed by Harsh Gupta

Photo 3.4 Founding leaders of Asian seismological commission, together with present leaders. From left to right: Harsh Gupta, Zhu Chuanzhen, Paramesh Banerjee, and Li Li, in 2018 during the 12th Assembly of ASC

3.3 ICSU Regional Office for Asia and Pacific

Despite all the technical and scientific developments, the impact of natural and human induced environmental hazards and disasters continue to increase with the passage of time. The important question is: *Why, despite advances in the natural and social sciences of hazards and disasters, do losses continue to increase*? Keeping this in mind International Council of Scientific Unions (ICSU) created a Regional Office for Asia and Pacific (ROAP). This region accounts for more than one-half of the world's population. About 80% human lives are lost in natural disasters, which occurs in this region. It was noted that the Tangshan, China earthquake of 1976 claimed ~ 250,000 human lives; The 1995 Kobe earthquake, Japan caused a financial loss more than US$150 billion; the Sumatra earthquake of 2004 and the Indian Ocean tsunami claimed over 250,000 human lives in south and south-east Asia and unprecedented financial losses. Two tropical storms in 1991 caused 130,000 deaths in Bangladesh; the 2002 land slide in Nepal displaced over 265,000 people. The Regional Office for Asia and Pacific (ROAP) chose to create a Planning Group on "Natural and Human Induced Environmental Hazards and Disasters in Asia and Pacific" as a priority area in September 2006. In the Board Meeting of ROAP at Tehran, Iran in March 2007, it was decided that the Planning Group of ROAP shall focus on three natural hazards, namely *FLOODS, EARTHQUAKES and LANDSLIDES* and I was asked to chair the group. The Planning Group had top experts in these fields from different geographical locations. The group had several meetings in 2007 and reviewed the work of several national and international bodies. An effort was made to collaborate with these bodies on appropriate topics and issues. In addition to the human and financial loses under ROAP jurisdiction, due to these hazards, several island countries were also brought under the program. It was found that the role of social scientists was very important due to trauma caused by these hazards, so the services of several social scientists were also called in. Two reports were brought out by the ROAP Planning Group.

(i) *Science Plan on Hazards and Disasters: Earthquakes, Floods and Landslides'* (https://council.science/wp-content/uploads/2017/05/ICSU-ROAP-Science-Plan-Hazards-and-Disasters.pdf)
(ii) *Science Plan on Hazards and Disasters: Special Vulnerability of Island* (https://council.science/publications/science-plan-on-hazards-and-disasters-special-vulnerability-of-islands-icsu-regional-office-for-asia-the-pacific/).

3.3.1 Socio-economic Aspects and Capacity Building

It was recognized that the available scientific knowledge is not being fully utilized in forecasting the hazard, estimating the risk, and the necessary outreach for the benefit of the public, particularly in a large part of the area covered by ROAP. Hence, the role of Social Sciences in understanding hazards and its mitigation was specifically stressed and recommended to bring the private sector to participate in alleviating the sufferings of the people. The need for enhancing capacity building disaster reduction and self-sustaining capacity at various levels for different hazards was pointed out. Several years have passed since this exercise was carried out. The issues raised and the possible solutions have gained more relevance with the passage of time.

3.4 Leading Asia Oceania Geosciences Society (AOGS)

AOGS was created in 2003 with a view to promote geosciences and its application for the benefit of humanity specifically for Asia and Oceania and an outreach to similar issues globally. The first annual convention was held at Singapore in 2004. The seventh convention was held at Hyderabad in 2010. I was one of the convenor for the Hyderabad session. In 2010 I was elected as the President of AOGS at the eighth convention held at Singapore (Photo 3.5).

Among the various responsibilities that I have had, being the President of the Asia Oceania Geosciences Society has been one of the most cherished and satisfying. The highlights being, overviewing the successful conduct of two Annual Meetings of AOGS at Taipei, Taiwan (2011) and Singapore (2012); inviting Nobel Laureates and other distinguished scientists globally to deliver prestigious AOGS Lectures; having the first Joint Assembly with American Geophysical Union (Western Pacific Geophysics Meeting) and adding a new Biogeosciences Section in 2012 to the existing six sections and an Interdisciplinary Working Group.

Photo 3.5 Prof. Y. T. Lee, Nobel laureate was facilitated during his visit to NGRI, Hyderabad while attending AOGS Convention

The mission of AOGS is to promote geophysical science for the benefit of humanity in Asia and Oceania is becoming more important with escalating catastrophes and changes that are affecting our environment and society. Asia and Oceania are the area most prone to natural hazards accounting for about 80% of human lives lost globally due to calamities of nature. The 11 March 2011 Mw 9.0 Japan earthquake and the resultant tsunami and the damage to the nuclear installations in Japan is one of the recent examples. An earthquake of this size was not anticipated in this part of the world. This calls for increased efforts in understanding the geological phenomenon and application of advances in natural and social sciences to better address hazard related issues. Responsive to these requirements, AOGS organized a very successful one whole day session dedicated to March 11, 2011 Japan earthquake at its 8th Annual Meeting at Taipei on August 12, 2011. AOGS took another important step at Taipei, in forward planning, and decided to hold the 10th and the 11th AOGS Annual meetings in the years 2013 and 2014 at Brisbane, Australia and Sapporo, Japan respectively. This provides ample opportunity to plan these meetings meticulously. AOGS has been very successful in providing an opportunity to showcase geoscientific products, techniques, publications to the world business community at its Annual Meetings. There were 25 stalls at Taipei, which were well visited. In my personal capacity, I very much enjoyed Springer's invitation that read "It is a pleasure to invite you to the "Author Meet and Greet" for the recently published "Encyclopedia of Solid Earth Geophysics" at the Springer booth A2, on August 9, 2011, 15.30–16.30 h with the editor Harsh Gupta in attendance". Interacting with several individuals about the genesis, contents and utility of this the two volume, 1500 + pages encyclopedia was indeed a delightful experience.

During 2011, while I was attending the IUGG General Assembly at Melbourne, Cheng-Hoon, from Meeting Matters (organizers of AOGS Annual Convention), invited me to visit Brisbane as there was an offer to host the next AOGS Annual Convention at Brisbane. My wife, Manju was happy to accompany me to Brisbane where we visited the proposed convention centre for holding the AOGS's future annual convention. We also visited Moreton Island, a most fabulous resort sea-resort.

From August 13 through 17, 2012, AOGS organized a Joint Assembly with the American Geophysical Union's (AGU) Western Pacific Geophysical Meeting (WPGM) at Sentosa Island Singapore. The 2012 has become a historic year in the annals of AOGS in its pursuit to expand regional and international collaboration in different disciplines of Geosciences for the benefit of mankind. One of the highlights of the 2012 Joint Assembly was the addition of the Bio-geosciences section. Over the past few years, Bio-geosciences have made a very rapid progress and its importance is well recognized. So, addition of this new section has been a step in the right direction. There was a wholehearted collaboration and cooperation with the AGU officials that led to a very successful and so far the largest attended meeting. Seeing the success of this joint venture, there is an interest expressed by AGU to have similar collaboration in the future. Another important happening at Singapore was the meeting of the seniors where several issues related with AOGS were discussed. The proposal to celebrate the 10th anniversary of AOGS at Brisbane, Australia in a be-fitting manner was accepted by all. The proposal to prepare a coffee table book of the 10 glorious years of AOGS was whole heartedly accepted. This coffee table book is the outcome of the deliberations in the meeting of the seniors.

During the 2016 Annual Convention of AOGS held at Beijing, China, I had the pleasure of receiving the Axford Medal: the highest honour bestowed by AOGS to an Earth scientist.

Here, I must mention that AOGS is a family of very well meaning individuals, who serve AOGS selflessly. The seniors are always available for advice and guidance and take pride in AOGS achievements and progress. We are lucky to have their guidance and continued support. The Secretariat of AOGS runs very efficiently and the response to a variety of requirements is excellent. The Annual Meeting books are prepared with great care and it is pleasure to use them.

Over the years, the importance of the geographical region covered by AOGS has increased in leaps and bounds. I believe AOGS has a very bright future and would play an important role in the well-being of the human race (Photos 3.6 and 3.7).

Photo 3.6 At the AOGS annual convention in Beijing, China in 2016. From the left: Minhan Dai, Robin Robertson, David Higgitt, Harsh Gupta, Manju Gupta, Guoxing Wu, Shail Mehta, Cheng-Hoon, and Jiaping Lee

Photo 3.7 Me delivering the Axford medal lecture at the 2016 AOGS annual convention in Beijing in 2016

3.5 International Union of Geodesy and Geophysics (IUGG)

I had been very deeply involved with the activities of the International Association of Seismology and Physics of the Earth's Interior (IASPEI) very early in my professional career. In 1995 I was elected to be an Executive Committee Member of IASPEI for 1991–1995 period, and at the end of the term I got elected to the Vice President position for the period 1995–1999. As a normal practice I was to be elevated as Senior Vice President at the end of my term but instead I was pleasantly surprised when Peter Wyllie, the President of IUGG who was on a visit to NGRI along with Bob Engdahl the Secretary General of IASPEI, offered to nominate me as the IUGG Bureau member, an elevation that gladdened me (Photo 3.8).

The IUGG 1999 General Assembly at Birmingham, U. K., would be remembered as NGRI made its first bid to host the 2003 IUGG General Assembly at Hyderabad in India. There was a very keen contest with Japan, and India narrowly lost the bid but there was something for us as I was elected a Bureau Member of IUGG. There were quite a few popular initiatives taken by me resulting in my re-election as an IUGG Bureau Member in the 2003 IUGG General Assembly held in Sapporo in 2003. All these years I was taking care of a few important topics that included recent damaging earthquakes and my efforts to involve young scientists in IUGG activities was very well appreciated. In the 2007 Perugia, Italy IUGG General Assembly, I was elected as the Vice President.

The IUGG General Assembly held at Melbourne from the 27 June to 8 July 2011 was a landmark occasion when our Encyclopedia of Solid Earth Geophysics was

Photo 3.8 29th IASPEI general assembly at Thessaloniki, Greece. Seated from right, Brian Kennett (President, IASPEI), Manju Gupta, Harsh Gupta, and Peter J. Wyllie (President, IUGG)

released in Melbourne. It was a very happy moment as all the 8 Editorial Board Members (Kusumita Arora, Anny Cazenave, Eric Robert Engdahl, Rainier Kind, Ajay Manglik, Sukanta Roy, Kalachand Sain and Seiya Uyeda) were present. Petra van Steenbergen, Springer's Coordinator from the publisher's side was also present for this grand book-release. The occasion became memorable as I was also elected as the President of IUGG for 2011–2015 quadrennial period and Prague, Czech Republic was chosen to be the venue of 2015 IUGG General Assembly. My tenure was very productive and path-breaking in many ways as several initiatives were taken up. Sincere efforts were made to engage younger generation of geodesists and geophysicists to be involved with IUGG activities. An IUGG Early Career Scientist Award was instituted. This Award honors early career scientists for their outstanding research in Earth and space sciences and for their international research cooperation. These Awards were presented by the IUGG President during the IUGG General Assemblies following the announcement of the awards. The award contains a certificate and covers travel expenses of the awardees to attend the General Assembly. The awardees were invited to give a talk at the General Assembly and the first awards were made during the IUGG GA at Prague. An IUGG Gold Medal was instituted for an individual for outstanding contributions in geodesy and geophysics. This prestigious IUGG recognition is for a scientist only once in every four years.

An IUGG Fellowship was instituted for individuals who have made exceptional contributions in international collaboration in geodesy or geophysics and attained eminence in the field of Earth and Space sciences. Honorary members, selected for IUGG Fellowship, are bestowed with a certificate of IUGG Fellowship by the President at the opening ceremony of the IUGG General Assembly. The number of fellowships granted is limited to ten every four years.

One of the IUGG's requirements was setting up of the IUGG Secretariat at a suitable location. During one of my visits to GFZ, Potsdam I met Prof. Dr. H. C. Reinhard Hüttl, Scientific Executive Director and Chairman of the Executive Board of GFZ on 17 June 2011. After some negotiations an agreement was reached and IUGG Secretariat was set up at the Helmholtz Center, Potsdam in 2013. Franz Kuglitsch was appointed as the Executive Secretary of IUGG/Assistant Secretary General of IUGG. Since, then the IUGG Secretariat is functional to the satisfaction of all concerned. A committee with C. Rizos, F. Kuglitsch, M. Mandea, A. Mackintosh and D. Giardini, was charged with the preparation of the IUGG Strategic Plan for 2016–2023. The Committee had been very active and had redefined the objectives, Mission and Vision Statements, Core Principals and Aspirations and Strategic Goals. The issue of Individual Membership of IUGG that was pending since long was addressed with the help of Association and National Members. Joint Assemblies of two or more Associations was also encouraged during this period.

Before I took over as the President in 2011, I had completed two terms as Member of the Scientific Planning & Review (CSPR) committee of ICSU and had contributed significantly. IUGG adheres to ICSU, and we brought out a resolution on Science on Disaster Risk during XXX General Assembly that was held in Rome from 27 to 30 September 2011. Nomination emerging from IUGG for the position of President-Elect of ICSU was successful when Gordon McBean from Canada became the President Elect of ICSU. The XXXI ICSU GA held at Auckland, New Zealand would be remembered for the in-depth discussions on the "Future Earth" initiative. At the end of the GA, Gordon McBean took over as the President of ICSU.

In the year 2012, a Union Commission on Climatic and Environmental Change (CCEC) was setup. The aim was to promote the scientific understanding of climate and environmental change, boost research for reducing uncertainties in climate and environmental models and encourage trans-disciplinary research on related topics. CCEC is expected to provide an all-Union perspective on the much-debated topic of climate and environmental change. Union Lectures are one of the most important activities of the IUGG General Assembly. We were lucky to have Noble Laureate Prof Y. T. Lee, the past president of ICSU delivering one of the Union Lectures.

During my term as the President of IUGG, several individuals provided support and advice making my task easier. Alik Ismail Zadeh, Secretary General of IUGG, needs to be thanked for constant advice and counsel. He had been all through very prompt and very knowledgeable about the IUGG affairs. He also carried the immense burden of day-to-day demands and operations of IUGG. Vice President Michael Sideris took care of his part of the responsibilities and provided support and advice. Treasurer Aksel Hansen had a long experience of managing finances of IUGG. He worked very well with the Finance Committee and had kept IUGG finances in an excellent shape. Members of the Bureau, namely, Isabelle Ansorge, Pierre Hubert and Kenji Satake came with different back-grounds and provided valuable inputs to run the IUGG secretariat. We found in Franz G. Kuglitsch a very young and energetic

Executive Secretary and Assistant Secretary General of IUGG, who was very quick in learning that energized the organization.

ICSU President Y. T. Lee (2011–2014) took immense interest in IUGG related activities and provided the necessary support. Whenever possible, he participated in IUGG related meetings. and his sensitivity, warmth, and the time he invested on these issues. I must acknowledge the support provided by the National Geophysical Research Institute, Hyderabad, and National Disaster Management Authority, New Delhi for running the IUGG President's Office.

As is the custom, as the immediate past President of IUGG, I chaired the Resolution Committee for the 2019 IUGG GA at Montreal, Canada, and the Nomination Committee of IUGG for the IUGG GA in Berlin in 2023. All these years it has been a pleasure being a part of IUGG, beginning in 1979 when I convened a session on 'Artificial Water Reservoir Induced Earthquakes' at IUGG GA in Canberra, Australia till the IUGG GA in Berlin, Germany in 2023. I was awarded the IASPEI Medal at Berlin in 2023. In all I have attended 12 IUGG GAs beginning from 1979 through 2023. This association has been most memorable and scientifically productive and rewarding (Photos 3.9, 3.10, 3.11, 3.12 and 3.13).

3.6 Compiling the Encyclopedia of Solid Earth Geophysics

Naturalis Historia is the oldest surviving encyclopedia prepared sometimes in the first century AD. Authored by Pliny the Elder, a Roman statesman, it has 37 chapters. The topics covered include history, geography, geology, and several other topics relevant in that era. Despite all the information available digitally, encyclopedias continue to be an in-depth source of knowledge. It all started during a visit to Cape Town, South Africa in January 2009, where I had gone to participate in the International Association of Seismology and Physics of the Earth Interior's (IASPEI) General Assembly. There I had a meeting scheduled with Petra van Steenbergen, a senior official of Springer Publishing Company, in connection with the Journal Geological Society of India which was co-published by Springer. During our discussions, Petra brought in the topic of publishing an Encyclopedia covering all the relevant topics of solid Earth Geophysics. Our discussions were spread over a week and ended up in my agreeing to writing a proposal for planning and editing the Encyclopedia of Solid Earth Geophysics. After returning to Hyderabad India, I had discussion with my younger colleagues Kusumita Arora, Ajay Manglik, Sukanta Roy and Kalachand Sain. All four of them were very enthusiastic about the project. Encouraged with these discussions, I then contacted Anny Cazenave from France, Eric Robert Engdahl from USA, Rainer Kind from Germany, and Seiya Uyeda from Japan. All these four very senior scientists and were globally known for their outstanding contributions in their chosen geophysics' disciplines. Their responses were very encouraging. Finally, with the help of all these eight colleagues, a proposal for the preparing the proposed encyclopedia was ready. The proposal was sent to Springer for the evaluation and concurrence. At that time, I had created an email account which was accessible to

Photo 3.9 IUGG GA In Prague, 2015. Harsh Gupta, President IUGG introducing Dr. Y. T. Lee Nobel-laureate for his union lecture

all the proposed 8 eight members of the Editorial Board. This was helpful as all the correspondences related with the proposed encyclopedia were accessible to everyone of us. A response was received from Springer in about three weeks' time. The proposal was found to be excellent. Only objection was to have 4 Editorial Board members from one institute in India. On the same afternoon, my four colleagues from the National Geophysical Research Institute came to meet me. They had all seen the response from Springer to our proposal and have had a meeting among themselves before coming to see me. They opined that I should withdraw their names from the Editorial Board, and they would be very happy to assist me in the preparation of the Encyclopedia without being a member of the Editorial Board. This suggestion was totally unacceptable to me. I told them that I shall do the encyclopedia only if Springer accepts to have all four of them on the Editorial Board. The same evening a mail was sent to Springer, thanking them for the appreciation of the proposal and conveying that in case having these four members on the Editorial Board was not acceptable to Springer, I would withdraw the proposal.

Photo 3.10 Prof. Sir Brian Hoskins receiving the first IUGG Gold Medal during IUGG GA in 2015

Photo 3.11 Early career scientist award to Iona Riipinen

Photo 3.12 Alik Ismael Zadeh (secretary general IUGG) with Harsh Gupta: after the IUGG GA in Prague in 2015

A response received from Springer after a few days accepting to have all four collogues from the National Geophysical Research Institute on the Editorial Board, gladdened all of us.

The work on the encyclopedia was started immediately. Around 250 topics were chosen to be covered in the encyclopedia. A systematic plan was devised and executed meticulously in identifying topics and probable authors to be approached. Everything from the design of the template and size of the articles were discussed and communicated to the authors. One main author for each article was chosen and the option of co-opting a coauthor was left to the main author. After obtaining firm commitments from the consenting authors, they were also informed about the topics by other authors that were closely related to avoid duplication. Every completed article was sent to a minimum of two reviewers. The review comments were communicated to the author, and I had the final responsibility to accept the revised and updated article by all the members of the Editorial Board. The main aim of the *Encyclopedia of Solid Earth Geophysics* was to serve as a comprehensive compendium containing all the key information on the important topics of Solid Earth Geophysics. The 200 + articles prepared for the encyclopedia covered both the established and new concepts of Geophysics across the sub-disciplines including Gravity, Geodesy, Geoelectricity, Geomagnetism, Seismology, Seismic, Deep Earth Interior and Processes, Plate Tectonics, Geothermic, Computational Methods, and related topics. Several Geotechnical Engineering and exceptional Exploration Geophysics topics were included for the sake of completeness. It may be mentioned that topics

Photo 3.13 Kenji Satake, President, IASPEI presenting IASPEI medal to Harsh Gupta, during IUGG GA 2023 in Berlin, Germany

pertaining near Earth environs were not included. It was felt that there is a scope of a separate encyclopedia for this vast field. While preparing the encyclopedia, it was kept in mind that basic understanding of Solid Earth Geophysics is essential for professionals dealing with allied subjects including Mining, Civil Engineering, Environmental Sciences, Exploration and Soft Ware Industry, as well as for NGOs working on a large-scale social agenda.

The 1550 pages, two volumes Encyclopedia of Solid Earth Geophysics having 217 article was ready by the summer of 2011 and was released during the XXV International Union of Geodesy and Geophysics General Assembly held at Melbourne during 27 June to 8 July 2011. All the members of the Editorial Board were present there. This was a very pleasant IUGG General Assembly for me as my birthday was celebrated in Melbourne on the 28 June and I was also elected to be President of the IUGG for the period 2011–2015. Later, on the publishers request I spent time on the Springer Booth at the Assembly signing the hard copies bought by individuals at the Booth (Photo 3.14).

The Encyclopedia was a grand success, was evident from the fact that every year there were 10,000 (ten thousand) or more downloads of the articles contained in the

Photo 3.14 ESEG editorial board (EB) members and other NGRI colleagues at IUGG 2011: From left: P. V. Sunder Raju, Bijendra Singh, Ravi Srivastava, Ajay Manglik (EB), V. P. Dimri, Ravi Kumar, Virendra Tewari, Y. J. Bhaskar Rao, R. K. Chadha, Harsh Gupta (EB), Sukanta Roy (EB) and Kalachand Sain (EB). In the front: Kusumita Arora (EB), Poonam Manglik and Manju Gupta

Encyclopedia. There was not a single error pointed out by any one all these years. Finally, by 2017, there were over 80,000 (eighty thousand) downloads. We were all very happy and celebrating the success of the Encyclopedia when Petra once again approached me in 2018 and expressed the wish to bring a new edition of the Encyclopedia as it had done very well and a lot of new developments had taken place in several branches of Solid Earth Geophysics. The offer was very interesting. I decided to have the same Editorial Board, as several of the 2011 edition articles were to be kept in the original form, being mostly of theoretical nature, several articles were to be updated by the earlier authors, or in case of their inability, to find replacements and over 40 new articles to be added. The problem was that the two members of the original Editorial Board informed that due to health and other personal reasons, they would not be able to actively participate in the preparation of the 2nd Edition. My 4 colleagues from the National Geophysical Research Institute had moved up to important positions and two of them had left NGRI for positions in other research establishments. All four of them promised to do the task at hand. After detailed consultations, a proposal for the second 2nd Edition of the encyclopedia was sent to Springer. In this proposal, the old Editorial Board was retained. After several articles from the 1st Edition with revisions as well as articles on new topics that were added, a final version was sent to Springer that was accepted after a thorough review process.

Photo 3.15 ESEG editorial board members at AGU 2019. From left: Rainer Kind, Anny Cazenave, Bob Engdahl and me

In the second edition I had to share the extra burden because of non-availability of two of my colleagues in the earlier editorial board. In early 2019 the COVID-19 pandemic hit the world. All of us were locked up in our respective homes. In a way, it helped to complete the preparation of the 2nd Edition, as we were no more spending time in traveling and all meetings were held online using Zoom and other platforms. The work was completed in late 2020 and the 2nd Edition was published in mid-2021. This edition has 1950 pages in two volumes and is doing very well. This is for the first time that a second edition of an Encyclopedia was published anywhere with same Editor and the Editorial Board (Photo 3.15).

Chapter 4
National Duty

4.1 India's Claim to Extra Exclusive Economic Zone

India has a very long coastline of more than 7500 km that allows the country to maximize its marine resources for commercial exploitation. The United Nations Convention on Law of the Seas (UNCLOS) came into existence in 1982 that defined the jurisdiction of coastal states over its maritime regimes. From the beginning India had played a constructive role in deliberations leading to UNCLOS's adoption in 1982 and had been a party to the convention since 1995. This convention is very important to India because it shares maritime boundaries with Pakistan and Maldives in the west, Sri Lanka in the south and Indonesia, Thailand, Myanmar, and Bangladesh in the east. Many of these boundaries are such that there are overlapping claims, especially when it comes to the extended continental shelf.

India had already invested heavily in the exploration of hydrocarbons, polymetallic nodules, cobalt and hydrothermal sulphide in the offshore region and deep seas around our coastline and needed opportune and accurate strategic measures to safeguard country's maritime boundaries. More and more hydrocarbon resources are being discovered in the offshore region and this requires a sound planning and strategy. According to UNCLOS the national jurisdiction of a coastal state extends from its territorial waters that starts from its coastal baseline up to 12 nautical miles to its Exclusive Economic Zone (EEZ) of up to 200 nautical miles of its continental shelf. So, any nation state with coastlines will have exclusive economic rights up to EEZ that include the sloping seabed that forms the part of the natural geological prolongation of its land mass.

When I took over as the Secretary to Government of India, Department of Ocean Development (DOD) in February 2001, I was aware of the provisions of article 76 of the UNCLOS that allows jurisdiction of a country over the living and non-living resources of the seabed in an *extended continental shelf* beyond the EEZ without having any claim on overlying waters. To lay claim over the extended area a country must make a submission, complete with the geological, oceanographic

H. K. Gupta, *Harsh K. Gupta*, Springer Biographies,
https://doi.org/10.1007/978-3-031-63561-8_4

and other details, to the Commission on the Limits of Continental Shelf (CLCS) up to a maximum distance of 350 nautical miles from its coastal baseline. The CLCS experts in the field of geology, geophysics and hydrography determine the scientific claims of the extent of the continental shelf by considering the data and information submitted by a member state.

It was realized that if India, with its vast coastline, can take up an exercise of geological and geophysical studies all around its off shore areas to determine the limits of its continental shelf in such a way that it maximises the area of the seabed for commercial exploitation, it would greatly benefit the country. With this in mind the Department of Ocean Development (DOD) embarked on an ambitious time targeted program to delineate the outer limits of the country's continental shelf beyond 200 nautical miles and submit its claims to CLCS.

The task was to collect geo-scientific data over 1.2 million square kilometers of offshore area all around the Indian coastline. The approach included a comprehensive desktop study of data that were available and additional data that were to be acquired, analyzed, interpreted, and documented for submission to CLCS. This required great synergy between the DOD and a host of other Ministries, Organizations, and Institutions to succeed in this national mission. The Goa-based autonomous institute of the DOD, National Centre for Antarctic and Ocean Research (NCAOR), that was established in 1998, was identified to lead this scientific endeavor and coordinate different task under one umbrella.

After several meeting and discussions, a mega maritime project on Indian Continental Shelf (LCS) was initiated to acquire data over 1.2 million square kilometers for which India's claim for the extended continental shelf beyond EEZ could be submitted. The entire data acquisition under the project was implemented during the year 2002–2004 in which over 40 Indian scientists and officials from 7 national organizations and institutes participated. In the sea, geophysical surveys were conducted along lines in a grid pattern covering a large area. The data acquisition began with the collection of marine geophysical data along 31,000-line km of multichannel seismic reflection, gravity, and magnetic data together with bathymetric information along 42 pre-determined profiles in India's EEZ in the Arabian sea, Bay of Bengal, and western part of the Andaman-Nicobar Islands.

Additionally, 100 Ocean Bottom Seismometers (OBS) were deployed in a first of its kind experiment with a retrieval rate of 92%. This allowed us to gather high quality wide-angle seismic reflection and refraction data at critical locations. The reflection data allowed us to derive the depth to the basement along different profiles through a process of iterative modeling of the two-way travel time of the acoustic signals passing through the water column and the sediments and sedimentary rocks above the basement. From this information we were able to estimate the depth to the seabed, the thickness of the sediments above the basement at thousands of discrete points along these profiles. Whereas the OBS data allowed us to constrain the velocity information from the reflection surveys and the magnetic and gravity data provided additional evidence on the nature of the basement.

Systematic single- and multi-beam bathymetric data for a precise determination of such parameters as the foot of the continental slope (FOS) and the 2500-m water depth were provided by the National Hydrographic Office (NHO), whose officials were also involved in the determination of the locations of the base of the slope and the FOS based on seabed morphology. The bathymetric coverage by the NHO in and off the country's EEZ was over 60,000 km.

The entire data processing, analysis, interpretation with rigorous quality checks was done indigenously at the CSIR-National Geophysical Research Institute, Hyderabad; NCAOR, Goa and CSIR-National Institute of Oceanography, Goa over a period of over two years. The NCAOR along with scientists from other national institutes carried out the integration of the analytical results and their documentation in accordance with the CLCS Guidelines. A thorough review of the documents was undertaken by a National Group of Experts who were responsible for obtaining approvals from the various Agencies of the Government of India before submission to CLCS.

India's Exclusive Economic Zone (EEZ) covers an area of about 2.37 million km^2 over which she has sovereign rights for the purpose of exploring and exploiting, conserving, and managing all living and non-living natural resources of the seabed and the waters above. The extended continental shelf claims of India under the first partial Submission are anticipated to add approximately 0.6 million square km of seabed area beyond the 200 nautical miles, over which she may exercise sovereign rights of the seabed and sub-seabed resources but not of the overlying waters (Fig. 4.1).

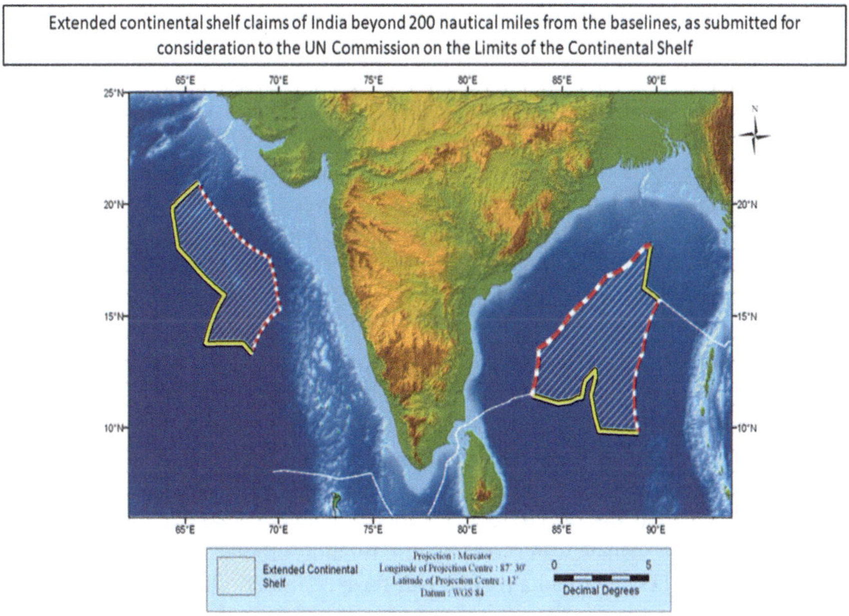

Fig. 4.1 0.6 million sq. km area that could be gained under legal continental shelf in the east and coast of India is shown by etched polygons

4.2 Setting up the First Low Temperature Thermal Desalination Plant in India

During my tenure as the Secretary to the Government of India, Department of Ocean Development, I was given a very challenging assignment that had great implications to the lives of the people in Lakshadweep islands of the Indian territory. I was very excited at the opportunity as it impacted the lives of a few thousands of people. The availability of the fresh water in Lakshadweep islands has been decreasing with the increase of population and several sources of water drying up. According to reliable estimates, in 1989, about 9,000 cubic meter of fresh water was available per person per year, globally. By the year 2000, the availability was reduced to 7800 cubic meters per person per year. It was estimated that by 2025 the availability shall drop to 5100 cubic meters per person per year. Despite 70% of the Earth's surface being covered with water only 3% of the total water available on Earth is fresh and potable. Of these 3%, only about 1% is available to human beings, the remaining being locked up as icecaps and glaciers in mountains and in Arctic and Antarctica regions. This 1% of water is being continuously re-cycled through the process of evaporation, rain, and other processes. Globally efforts are being made to desalinate the sea water using Reverse Osmosis, a popular methodology. However, the disposal of the residual brine concentrate is a major problem.

The Low Temperature Thermal Desalination (LTTD) involves taking the warm surface sea water in a flash chamber. It is flash evaporated under appropriate low pressure and the vapors are condensed by using cold water brought from suitable depth from the sea. The LTTD is a very suitable technique for tropical countries located between 20° N to 20° S latitudes. The principle of LTTD is known for quite some time. However, no major effort was made till 2004 for a mass production of potable water using this technique. The National Institute of Ocean Technology (NIOT), Chennai, an institute under DOD was involved to set up a pilot plant of 5000 L/day capacity to test the efficacy of the method. The plant was set up and successfully operated in Chennai. This encouraged us to look for a suitable location for setting up of 100,000 L/day capacity plant. The plant was proposed to be set up on land near the coast where the sea slope was very steep so that the cold water from the depth can be utilized. One such location was found in the Kavaratti Island in Lakshadweep group of islands in the Arabian Sea. Kavaratti hosts about 10,000 residents and is located 200 km west of Kochi, a major city on the west coast of India in Kerala. There is a good connectivity of Kavaratti to Kochi through air and sea routes. The suitability of Kavaratti for setting up of a LTTD plant was governed by the fact that within 400 m of Kavaratti coastline, the sea deepens to 350 m, and the sea water temperature drops to 12 °C, while the surface water temperature remains around 24 °C round the year. The Kavaratti LTTD plant had a rectangular flash chamber of about 34 cubic meter capacity. After creating a vacuum, surface sea water is flashed into the chamber. The vapors generated are taken to a cylindrical shape condenser having a length of 7.5 m and a diameter of 1.5 m. The water vapor is condensed into fresh water using deep sea cold water being brought from a depth of 350 m using an HDPE pipe. The shore end

Table 4.1 Specifications of the Kavaratti LTTD plant [after Kathiroli et al. (2006), Low temperature thermal desalination plant at Kavaratti, Lakshadweep, J-Geo Soc India, 67, 820–822)]

Fresh-water generation rate	100,000 L/day
Warm-water temperature	28 °C
Cold-water temperature	13 °C
Cold-water flow rate	180 kg s^{-1}
Warm-water flow rate	145 kg s^{-1}
Cold-water intake	From 350 m water depth
Vacuum maintained	Low vacuum level

of the HDPE pipe is connected to sump. The sump has two compartments for cold and warm water. Using a coarse screen to filter foreign particles. The specifications of the Kavaratti LTTD plant are provided Table 4.1. The quality of the fresh water generated is consistent with the requirements of Bureau of Indian Standards as well as the World Health Organization.

The LTTD plant has been operating at Kavaratti uninterrupted since 2005. The number of cases reported in the hospitals in Kavaratti have dropped to less than one-half of the pre-LTTD plant era. Most patients had problems related water. Availability of 10 L of potable water per person per day has made the difference. The cost of the plant was in 2005 was about 1 million US $. It was then estimated that if the plant runs successfully for 10 years the cost of potable water would be about 60 L per US $. Another benefit of setting up of this plant has been the increase of fish catch, as the water brought from deep sea is full of nutrients and when let out, it helps growth of fish. Seeing the success of Kavaratti LTTD, 5 more LTTD plants have been installed in Kavaratti at Agati, Minicoy, Kalpeni, Amini and Kadmat. It is very interesting to note that India succeeded in setting up of the first 100,000 capacity LTTD plant globally. Its success has led to setting up of similar plants elsewhere in Lakshadweep group of Islands (Figs. 4.2, 4.3 and 4.4).

4.3 Establishing Indian Tsunami Early Warning System for the Indian Ocean

On 26 December 2004 people returning from morning walk and fishermen residing along the eastern coast of India from Vishakhapatnam in Andhra Pradesh to Kanyakumari in Tamil Nadu observed a strange phenomenon of the sea receding few hundred meters inside exposing parts of the coastal shelf. What they were observing was a precursor to the incoming gigantic tsunami waves generated by one of the deadliest undersea earthquakes in modern times. Oblivious to the lurking danger, people walked in to pick shells, fish and other undersea objects were swept away with the arrival of the first of the tsunami waves that reached the shore.

That day a Mw 9.2 magnitude earthquake off the coast of Banda Aceh, northern Sumatra broke open and displaced the ocean floor by about 18 m causing large

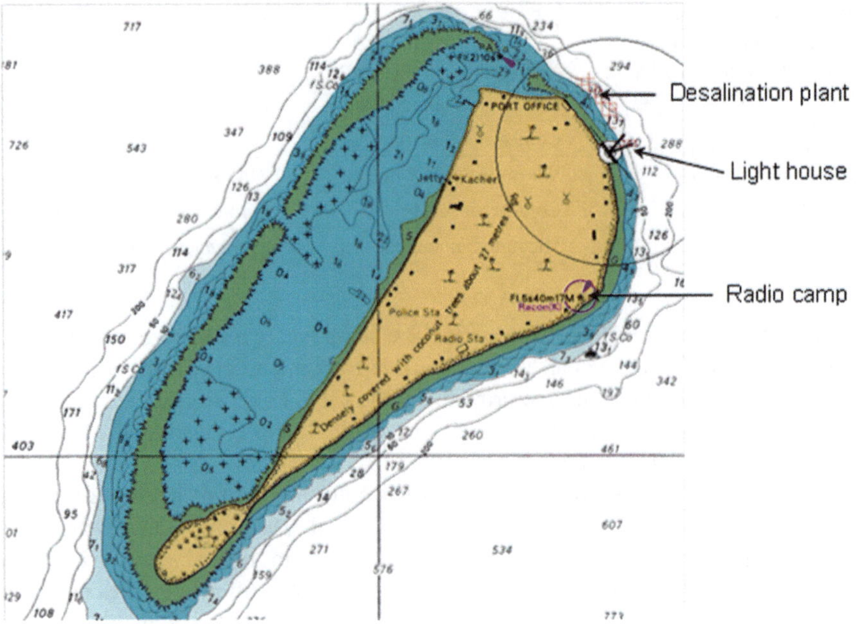

Fig. 4.2 The figure shows Kavaratti Island, located 200 km west of Kochi, India, and the bathymetry around it. The island is 5.6 km in length and the maximum width is 1.2 km. water depth contours are shown in meters. Spot values indicate water depth in meters. Coral reefs (indicated by crosses) are located to the west of the island. The LTTD plant is located near the northeastern margin of the island (Gupta and Roy, 2008)

scale displacement of water generating tsunami waves. Once the tsunami waves were generated it travelled thousands of kilometres in the open Indian Ocean and affected the coastal region of more than 14 countries with most devastating impact in Indonesia, Thailand, India, and Sri Lanka. Since, the tsunami waves have long wave lengths of 200–250 km, their height in the open sea hardly reaches sub-meter levels and cannot be distinguished by people travelling on a ship in an open sea. The tsunami waves gain height and become slower as they move to coast due to shallowing of the sea.

The average distance from Banda Aceh to the Indian eastern coast is about 1800 km and it took about 2 h and 30 min for the tsunami waves to reach the India's mainland shores. However, Andaman and Nicobar Islands being located much closer to the tsunami source, the tsunami reached after a couple of tens of minutes and caused heavy damage and casualties. In the mainland India, Tamil Nadu and Andhra Pradesh were the worst affected. Chennai city bore the maximum brunt where tens of morning joggers and fishermen perished. More than 25,000 lives were lost in India and nearly 300,000 lives in Asian countries, Indonesia taking the heaviest toll of more than 175,000 people.

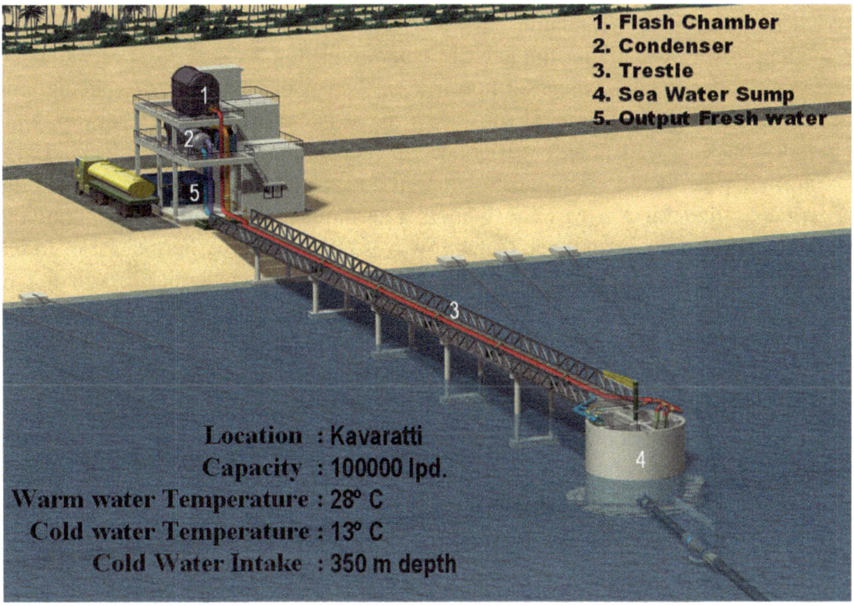

Fig. 4.3 Conceptual layout of the LTTD plant at Kavaratti island (Gupta and Roy 2008)

Fig. 4.4 A picture of the 100,000 l/day capacity LTTD plant in operation at Kavaratti [Kathiroli et al., (2006) Low temperature thermal desalination plant at Kavaratti, Lakshadweep, J-Geo Soc India, 67, 820–822)]

Till then, the word 'Tsunami' was not very familiar to general public and also to media persons. Soon after the news of tsunami hitting the east coast of India and Andaman Nicobar Islands started pouring in, there were all kinds of speculations and stories conveying a general sense that the water of the sea was boiling. Half a century back twice tsunami waves had hit the Indian shores in 1941 and 1945. The former was due an earthquake of magnitude 8.1 in Andaman that affected the eastern coast and the later was due to another earthquake of magnitude 8.1 along the Makaran coast in Pakistan when tsunami waves reached the western coast of India. In both these cases tsunami waves were insignificant and generally less than 1 m height, whereas the 2004 tsunami waves reached a maximum height of more than 5 m at Nagapattinam in Tamil Nadu with lateral inundation of up to 1 km. Most of the loss of life and damage to property was within 500 m of the shoreline and the local coastal topography played an important role in the inundation process.

There was a grave concern in the government because of the magnitude of the tragedy and also a great concern as two of the India's Nuclear Power Plants namely Kalpakkam and Kudankulam are located along the east and southern coast of India. A survey later showed that tsunami wave at Kalpakkam was about 4 m and inundation about 800 m. Relatively, Kudankulam in Tamil Nadu had lesser waves heights because of the presence of Sri Lankan landmass which took the first brunt of the tsunami waves, being directly in line with focus of the tsunami wave direction.

In the Pacific Ocean where tsunami are common, the Pacific Tsunami Warning Centre (PTWC) issues bulletins and early warnings on tsunami to countries in the Pacific Ocean regions. The PTWC was established in 1949, following the 1946 Aleutian Island earthquake and a tsunami that resulted in 165 casualties in Hawaii and Alaska regions of USA. The Japan Meteorological Agency (JMA) is responsible to provide bulletins and warning in the Pacific Ocean and coordinate with the international tsunami warning systems. It is noteworthy that in the 20th Century around 800 small and large tsunami have occurred in the Pacific Ocean compared to just 3 in the Indian Ocean.

Since, tsunami is a rare occurrence in the Indian Ocean region, there were no efforts to establish a tsunami warning system. After the 2004 Indian Ocean tsunami several landmark decisions were taken at the highest level in meetings chaired by the Prime Minister, Minister for Science and Technology, Cabinet Secretary and setting up of a Tsunami Warning System for the Indian Ocean became a priority.

"Cometh the hour, cometh the man". With the right credentials and being at the right place as the Secretary of the Department of Ocean Development and at the right time, I became the natural choice of the decision makers at the highest level to lead India's efforts to establish a Tsunami Early Warning System for the Indian Ocean region. This was to help India as well as all other Indian Ocean rim-countries with timely warnings of this ocean disaster and take the leadership role. A few of the international Tsunami Centre Operators came up with offers to set up a tsunami warning system on a turnkey basis costing approximately INR 1000 crores (~ US $ 160 million in 2005). In the absence of a dedicated tsunami warning facility for the Indian Ocean, often, advisories issued by the Japan Meteorological Agency (JMA), and the Pacific Tsunami Warning Centre (PTWC) located in Hawaii were acted

upon. Sometimes these turned out to be false alarms. After the 26 December 2004 earthquake, there were incidences where tsunami advisories were issued and millions of residents in Tamil Nadu on the east coast of India moved away, and later it was found to be a false alarm. This caused immense inconvenience to residents. The entire 40,000 km Circum-Pacific earthquake belt passing through several countries could generate tsunamigenic earthquakes. The time taken for tsunami to travel to populated coastal areas at times is just a few minutes as the tsunamigenic earthquakes occur very close to these inhabited areas. While the earthquake is located within a minute or so of its occurrence, whether it has generated a tsunami is not known till the tsunami wave reaches the nearby located tide gauges. Therefore, the *modus operandi* by JMA and PTWC is to issue a tsunami advisory soon after the occurrence of the possible tsunamigenic earthquake and later withdraw the advisory if no tsunami is generated. People living in the vicinity of the Pacific Ocean coast are familiar with this approach.

Detailed investigations revealed that as far as the Indian Ocean is concerned, there are only two sources for the tsunamigenic earthquakes: a stretch of about 4000 km from Sumatra to Andaman region, and an area of about 500 km radius off Makaran Coast in the Arabian Sea. Tsunami generated by the Sumatra–Andaman Island source takes about 15–30 min to reach Andaman & Nicobar Islands, and over 90–120 min to reach different parts of the east coast of India. As far as the Makaran source is concerned, the tsunami waves reach the Indian west coast in about 70–90 min and are not very large due to the bathymetry structure. The west coast has a longer continental shelf where the sea is shallow, approximately 20 m deep and extends to 100 km or more westward into the ocean. Therefore, the tsunami gets weakened by the time it reaches cities like Mumbai, Goa and further south as was observed in the 1945 tide gauge records. It is therefore necessary to plant Ocean Bottom Pressure Recorders in the immediate vicinity of these two tsunamigenic sources in the Indian Ocean, so that, after the occurrence of a magnitude 7 or larger earthquake in either of the two locales, whether a tsunami is generated or not is detected by the ocean bottom pressure recorders within a few minutes of the occurrence of the earthquake and suitable advisory could be issued.

These finding were shared with the Minister for Science and Technology, Mr. Kapil Sibal. He was convinced with the logic and informed the Prime Minister. It was decided not to accept the foreign offers of setting up of a tsunami warning system for the Indian Ocean, and the Department of Ocean Development was charged with the responsibility of setting up of a suitable tsunami warning facility for the Indian Ocean.

The International Oceanographic Commission (IOC) called for a meeting in Paris, France in March 2005 to discuss the issues arising out of the Sumatra earthquake tsunami and how to move forward. The IOC has a membership of over 150 countries and the meetings are held in Parliamentary style. India was also invited being one of the main stakeholders, and I was asked by the Government of India to lead an Indian Delegation of scientists that included Dr. K. Radhakrishnan, Director, INCOIS, Hyderabad, where the Indian Ocean Tsunami Warning Center was to be located. Ms. Bhaswati Mukherjee, Permanent Delegate of India to UNESCO was our diplomatic support. There were intense and detailed discussions on how to go about and provide the Indian Ocean countries suitable tsunami advisories to avoid human loss of life in future.

In my presentation I raised the issue of false alarms and the avoidable inconvenience caused by them. It was also pointed out in my presentation that there are only two regions in the entire Indian Ocean which can host tsunamigenic earthquakes, namely the ~ 4000 km stretch from Sumatra to Andaman, and the area off Makaran Coast in the Arabian Sea. I suggested the approach that we were working on, where ocean bottom pressure recorders were to be placed in the vicinity of the two sources of tsunamigenic earthquakes to detect tsunamis soon after their generation. As I was finishing my presentation, a handwritten note from Ms. Mukherjee was passed on to me that stated:

> You should insist with the Chairman to take a few questions, after India finishes. There is a great interest in your presentation, and you were within the time limit.

Her suggestion was a good one and I followed. My presentation had generated a lot of discussion on the views expressed by me. Arguments for and against my proposal for the alternate model were hotly debated. There was no general agreement on the view that there are only two tsunamigenic earthquake sources in the Indian Ocean. As doubts were raised and no consensus was arrived at, I requested the Executive Secretary of IOC, Dr. Patricio Bernal to have a discussion after a fortnight on the same issue, when the participants could find another source of tsunamigenic earthquakes in addition to the two sources identified in my presentation. He saw merit in my arguments. At the end of the deliberations, the following Communique was issued:

> Noting it is important to improve the science of issuing tsunami warnings to reduce false alarms given the inordinate inconvenience and disruptions to normal life caused by false alarms, especially, given the high population densities and intensive operations in coastal areas in the Indian Ocean, and also to continuously improve forecasting…

I was extremely pleased that the IOC took my view seriously and I was able to put forward the viewpoint from the Indian perspective. The delegation returned to India with a sense of achievement.

After a fortnight, on 28 March 2005 I was returning to Delhi by the last flight from Chennai in the evening. Just when I landed and switched on my phone, I got a call from my wife informing me that there were numerous frantic phone calls from media and TV news channels enquiring about the tsunami advisory issued by PTWC as an earthquake of 8.6 magnitude occurred close the 26 December 2004 earthquake epicentral region. This earthquake was later named as the Nias earthquake. There was a panic as people were asked to vacate areas near the seashore. They were all eager to know whether a tsunami is generated or not as the memories of the Indian Ocean tsunami three months back were still fresh in their minds. I asked them for a few minutes so that I could gather some information and look at some data on this earthquake.

While looking into the seismological data from the records of Indian National Network operated by the India Meteorological Department, New Delhi I called up my contacts in Djakarta, Indonesia and sought information on the earthquake and any tsunami warning issued. It was informed that a large earthquake of 8.6 magnitude had occurred offshore of West Sumatra, Indonesia and had caused substantial shaking, damage, and land level changes between Simeulue Island in the north and the Batu Islands in the south. The earthquake generated only a local tsunami and there was no threat for the Indian Coastal region. Consequently, I informed the media and the TV channels that there is no tsunami threat to the Indian coast because of this earthquake. However, the damage was already done as close to 7 million people had moved away from the coastal areas along the east coast of India after the tsunami advisory was issued, earlier. This caused immense difficulties to a large population and created unnecessary panic. The incident further strengthened my resolve to set up our own system of tsunami warning in the Indian Ocean with a missionary zeal so that no such false alarms are issued.

I took up the case of Nias earthquake and again pitched for the necessity of developing a false alarm proof Indian Ocean tsunami warning system at the 2nd International Coordination Meeting for the Development of an Indian Ocean Tsunami Warning and Mitigation System that held a month later at Grand Baie, Mauritius during 14–16 April 2005. In the 1st meeting in Paris several scientists were not convinced of my views and were resisting our efforts to have a system totally implemented indigenously. But the Nias earthquake in March 2005 in Sumatra that did not generate huge tsunami waves came in support of my arguments that the situation in the Indian Ocean is different with that of the Circum-Pacific. In the end, a declaration was made that was in line with the arguments reflecting India's perspective. The relevant points of the Mauritian Declaration at number 5 and 6 are reproduced here:

4.3.1 Mauritius Declaration (2005)

5. *...Recognize the unique tectonic plate structure of the Indian Ocean, and that there are primarily two tsunamigenic sources that could affect the coastlines of the Indian Ocean, namely the Indonesian seismic zone and its extensions, about 4000 km in length and the Makaran source;*
6. *...Welcome the plans and intensions of Australia, India, Malaysia and Thailand to develop their national capability to detect, analyze and provide timely warning generated along the Indonesian seismic zone and its extension as well as the plans of India, Iran and Pakistan to cover the Makaran coast.*

After the Mauritius meeting things started moving very quickly. A concrete project was finalized after discussions with scientists from several organizations within the country based on the data we had collected immediately after the 26 December 2004 Indian Ocean tsunami. I also kept the office of the Prime Minister, Minister for Science and Technology, Cabinet Secretary and all other major functionaries of the government informed about the developments. Finally, when we finalized the proposal, the Government of India was approached for necessary approvals. The

Government of India at the highest level expressed total confidence in the proposal and issued a notification that India would go ahead with setting up of the Indian Ocean Tsunami Warning System on her own and under my leadership. It was yet another great moment in my life to establish a system that will not only serve the coutry but also the population all along the Indian Ocen-rim countries from tsunami hazard.

The President of India, Dr. A. P. J. Abdul Kalam, a scientist of international repute called me immediately and expressed his satisfaction and happiness that a scientist is heading this great Indian initiative. He called me several times to his office, later for inputs on the progress of the project. To make people know about my proposal and give confidence to the nation I wrote about the details of Tsunami Early Warning System and boldly made a commitment that was published in *Episodes*, a scientific journal in the March Issue in 2005, immediately after the Paris meeting. I reproduce excerpts of that commitment here,

> The Project will be completed within 30 months at a cost of US $ 30 million. With the commencement of the project in February 2005, operational warning of Tsunami and Storm Surges along with their likely landfall and inundation could be provided from September 2007.
>
> It may also be noted that tsunamigenic zones which can produce tsunamis for the Indian coasts are also the principal source of tsunamis for the rest of the Indian Ocean rim countries. We shall be providing products of our work to all the countries and center of research, through the mechanism of Indian Ocean Goose or other suitable mechanism. India shall also be interacting and coordinating with other Tsunami warning and Research Centers globally.

As planned it all began with the installation of 'Ocean Bottom Pressure Recorders' in the immediate vicinity of the two tsunamigenic earthquake zones. Installation of these instruments was crucial and key to avoid issuing false alarms. In the open seas waves heights of a few meters occur all the time due to atmosphere-hydrosphere interaction. However, at depths, these waves die off and the sea is very calm and quiet. The primary mechanism of tsunami generation is the displacement of water due to earthquake faulting under the sea causing vertical upliftment of water that falls down under the force of gravity generating a tsunami wave. This displaced column of water oscillates and spreads out as a tsunami wave travelling thousands of kilometers in the open sea. There are also a few other secondary mechanisms like large scale submarine landslides which can cause tsunami. So, if pressure sensors are placed in the vicinity of the tsunamigenic earthquake zone, it will decipher the occurrence of a tsunami after the tsunami. The amplitude and other characteristics of the tsunami are also estimated from the recordings of the ocean bottom pressure recorders.

In addition to installation of pressure sensors at the ocean bottom, several tide guages that monitored sea level changes continuously were also installed as an additional parameter. Radar-based surface current and wave monitoring in real time was started. Additional seismological observatores were established at critical locations adding to the existing seismic networks in the country to record earthquakes with great accuracy from these two tsunamic earthquake sources. Parallel efforts were started to numerically model tsunami wave propagation in the Indian ocean to estimate waves heights and inundation along along the Indian coast. The work for the

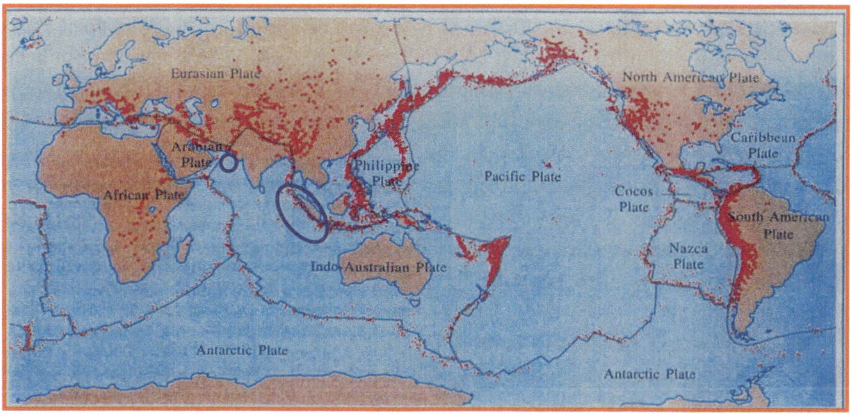

Fig. 4.5 Diagram showing the distribution of earthquakes and major plate boundaries. It may be noted that globally, more than 75% of earthquake energy is released in the circum-Pacific belt, about 20% in the Alpine-Himalayan belt, and remaining

construction of a Tsunami Early Warning Centre in the campus of INCOIS, Hyderabad was initiated and completed in a record time. The entire data sets from all these sensors was connected to this centre where the monitoring was done on a 24 × 7 basis.

It was sense of great personal achievement when the Indian Tsunami Early Warning System became fully operational in August 2007 in the campus of INCOIS, Hyderabad, one month before my commitment made to the nation in 2005. Dr. Shailesh Nayak who took over as the Director, INCOIS was of immense help in fulfilling this dream. As if to test the efficacy of the system to perform as promised an undersea earthquake of magnitude 8.4 occurred in southern Sumatra on 12 September 2007. It was a major earthquake felt over a large area with reports of tsunami waves reaching 8–10 feet at Padang, Indonesia.

12th September 2007 became a *Red-Letter Day* in the history of Indian scientific achievements when the first successful and accurate tsunami advisory was issued for the Indian coast by the infant Tsunami Early Warning System. The Minister of Earth Sciences, Mr. Kapil Sibal dedicated the system to the nation on 15 October 2007 in the presence of the Chief Minister of Andhra Pradesh, Mr. Y. S. Rajasekhar Reddy (Fig. 4.5).

4.4 Presiding Indian Science Congress

Science Congress is the biggest annual science event in India. The first meeting of the Indian Science Congress was held at Kolkata in January 1914 attended by over 100 scientists from all over the country. One of the aims of the Indian Science Congress is to create scientific temper in the country and bring awareness in the rural areas about the benefits of studying science that leads to rational thinking that is central to the scientific development of a nation. The Indian Science Congress gained popularity and became the largest convention for scientists in all disciplines and the number of participants grew steadily, exceeding 5000. However, no Science Congress was ever held in the state of Kerala till 1989. As the Vice Chancellor of the Cochin University of Science and Technology, Cochin, I had hosted the first ever Science Congress in the State of Kerala in the year 1990. It was a grand success. Prof Yash Pal was the General President. It was inaugurated by the then Prime Minister of India, Mr. V. P. Singh and was very well received by the public and academics.

In the year 2006 I was elected as the General President to host the 94th session of the Indian Science Congress to be held in January 2007. After looking into several options to hold this mega event it was decided to host it at the Annamalai University in Chidambaram, Tamil Nadu.

I was stationed at NGRI during the preparatory year 2006 and had a dedicated team of scientists to support me. The Vice Chancellor had extended all the support and was very excited to have such a prestigious event in their campus. The Indian Science Congress is held every year in the first week of January and the Prime Minister of the country inaugurates it. This is largest scientific meeting in India with more than 5000 participants all over the country and hundreds of scientists from all over the world. The scientific sessions cover all disciplines of natural sciences. Once again, I found myself in the right place and occasion being an Earth scientist. The United Nations General Assembly had declared 2007 as the International Year of Planet Earth (IYPE) to increase awareness of the importance of Earth sciences for the advancement and sustainable development. UNESCO was designated as the lead agency and the Year's activities spanned over three years starting from 2007 to 2009.

Looking to put India on the world map once again it was decided to adopt the theme *"Planet Earth"* for the 94th session of the Indian Science Congress in 2007 coinciding with the IYPE. This gave an international visibility to the Indian event attracting several Nobel Laureates and scientists from abroad, a very proud moment for all Indians who attended the meeting although the venue was not directly connected to an international airport. One must travel by road from Chennai to Chidambaram by road and it took around 4 h. Dr. Manmohan Singh, Prime Minister of India inaugurated the Indian Science Congress on 3 January 2007 at Annamalai University and stressed the importance of Earth Sciences and the need to stop degradation of our planet. He urged the scientific community to bring awareness to preserve our planet, the only one to support life and not self-destruct ourselves with mindless activities of development. He strongly advocated for the need to have sustainable development. Mr. Kapil Sibal, Minister for Science and Technology and Ministry of Earth Sciences underlined Government of India's commitment to integrate all earth science related activities under one Ministry and he cited creation of a separate Ministry of Earth Science in 2006 as an example. In my Presidential address I made 11 specific recommendations that are required for sustainable development as pointed out by the Prime Minister of India.

Dr. A. P. J. Abdul Kalam, the President of India while launching the IYPE (International Year of Planet earth) event at the Science Congress addressed the scientists

Photo 4.1 Indian Science Congress with Prime Minister Dr. Manmohan Singh, Chief Guest for the inauguration

of the country and asked them to work on a concept he described as *"PURA"*. The acronym PURA is Provision of Urban Amenities to Rural Areas. Dr. Kalam coming from a rural background had a ringside view of the problems of the rural people and he was working passionately to improve their living conditions. Dr. Kalam's love for teaching and interacting with school children is a subject of folklore in India. Knowing his love for children a separate session with school children was held that was televised all over India. Before leaving the venue of the Science Congress, he formally launched Planet Earth related activities in India by releasing symbolic balloons on the occasion.

This Science Congress went into history with very participation of scientists from abroad including Nobel Laureates. It may be noted that in the year 1947, Pandit Jawahar Lal Nehru was the General President of Science Congress, and in the year 2013, when the Indian Science Congress was celebrating 100th year, Dr. Manmohan Singh was elected as the General President of Science Congress. After the event I felt a great job done and prepared myself for the next assignment (Photos 4.1, 4.2, 4.3, and 4.4).

Photo 4.2 Indian Science Congress, Annamalai University Campus. Dr. Thomas Rosswall, Director of ICSU, Paris (right) and Andrzej Kijko from South Africa in front of big hoarding

Photo 4.3 Indian Science Congress. With President of India, Dr. A. P. J. Abdul Kalam releasing balloons to declare beginning of the International Year of Planet Earth for India

Photo 4.4 Indian Science Congress. Prime Minister of India, Dr. Man Mohan Singh meeting Dr. Hari Narain and other top scientists in the country

Chapter 5
Dream Projects Initiated

5.1 7-km Deep Borehole in Koyna

After completing my tenure at DOD, New Delhi and seeing its transition into a full-fledged Ministry of Earth Sciences, I once again came back to NGRI, my home in Hyderabad as a Raja Ramanna Fellow. This was the most exciting phase for me as there was the time to work on my dream project of understanding genesis of earthquakes processes. I had been thinking of an opportunity to work on this subject since the occurrence of the devastating Koyna earthquake in Maharashtra on 10 December 1967. All these years while busy with national and international assignments I had always kept my touch with the subject and suggested the mechanism of triggered earthquakes in the Koyna and Warna reservoirs region in Western Ghats.

My research on common characteristics of triggered earthquakes in the Koyna region published in the form of papers and books in the reputed international and national scientific journals were widely debated. After the initial reluctance this research work gained international acceptance and it was thought appropriate to proceed further and go deeper into the understanding of earthquake genesis. There was fascination that if we can drill a deep borehole in the earthquake source region of Koyna, we may probably catch an earthquake *Live*. The idea was simple to map the entire timeline of processes from the initial building up of stresses to the final snap when dislocation happens and earthquake occurs. If what we understood so far about the physics of rock failure from laboratory experiments, then changes in certain parameters should be reflected in the earthquake source region. The fact that earthquakes in the Koyna and Warna region have been continuing for the past more than five decades, makes this location a suitable site to be used as a natural *Field Laboratory* for earthquake genesis studies. Since, earthquakes are a random phenomenon finding a location where occurrence of earthquakes is assured in one's life time is also random, by logic. Koyna fits the bill in this regard with assured supply of earthquakes every year and if a concerted effort is launched where one

H. K. Gupta, *Harsh K. Gupta*, Springer Biographies,
https://doi.org/10.1007/978-3-031-63561-8_5

can literally look into the earthquake source region then there is strong possibility to observe last minute changes in the parameters of the rocks before its failure.

The amount of work that had been carried out on triggered earthquakes in Koyna for the last six decades has improved our understanding of earthquake source region. The greatest advantage of the region is that the earthquake source volume is tightly constrained in a very localised region where earthquakes repeat itself in a regular annual pattern. There is a rich account of scientific research available as an a priori knowledge that adds to our advantage. Because of such enormous data available I had made a few attempts to forecast moderate size earthquakes in the region with an excellent success rate, presented earlier in this book. I had been advocating this concept at several platforms on the national and international level for drilling a deep borehole to a depth of 7 km in the seismically active Koyna-Warna region that defines the heart of the earthquake source volume with an objective to catch a *Live* earthquake.

During a visit to Geo Forschungs Zentrum (GFZ), Potsdam, Germany I spent time with the then Director of the International Continental Drilling Program, Pro. Dr. Rolf Emmermann on 30 September 2010. We had a long chat and he was convinced that Koyna, India is a suitable site to put a deep borehole going through the earthquake source zone to better understand the process of earthquake nucleation, an observation that is missing globally, and no available earthquake model has been tested in the field. This idea generated a great interest at the Ministry of Earth Sciences that culminated their readiness in funding such a project at a national level as it was addressing an important and fundamental Earth science issue. The synergy that was seen during several meetings of different organization in the country gave great hopes to me to fulfil this dream meant for the welfare of the mankind.

At the international level these efforts bore fruits when International Continental Scientific Drilling Program Group in Potsdam, Germany agreed to co-sponsor an International Workshop on Deep *Scientific Drilling to Study Reservoir Triggered Earthquakes in Koyna India* with Ministry of Earth Sciences, Government of India. The Workshop was held from 21–25 March 2011 at NGRI, Hyderabad, and Koyna, where 81 participants including 26 from USA, Japan, Germany, France, Italy, Poland, Taiwan, Canada and New Zealand attended.

The principal objective of the Workshop was to fully review the plan for deep drilling down to earthquakes focal depths of ~ 7 km at Koyna and prepare a roadmap for implementation. The first two days of the Workshop were held at NGRI, Hyderabad, followed by a field visit to Koyna-Warna area for onsite discussions and familiarization with field conditions. Participants from abroad included seismologists and experts associated with active fault zone drilling projects worldwide such as the San Andreas Fault Observatory at 5 km Depth (SAFOD) in western USA, the Chi-Chi earthquake of Taiwan, the Kobe earthquake of Japan, the Gulf of Corinth in Greece and the Latur earthquake in India.

55 Indian scientists representing major Earth Science Organizations, Institutes and universities in the country such as the MoES, NGRI, Geological Survey of India, Atomic Minerals Directorate for Exploration and Research, Oil and Natural Gas Corporation, North-East Institute of Science and Technology, India Meteorological

Department, Institute for Seismological Research, Wadia Institute of Himalayan Geology, National Institute of Rock Mechanics, Central Water and Power Research Station, National Centre for Antarctic and Ocean Research, Pune University, Sivaji University (Kolhapur), M.S. University of Baroda, participated.

The major success of the Workshop was the unanimous agreement by all national and international participants that Koyna is an outstanding world geological site to plan a deep borehole experiment for earthquake studies. Suggestions were made to conduct the following investigations before drilling the borehole. Some of them are:

- Improving the location of the hypocenters in the region by adding more seismic stations
- Conducting airborne gravity and magnetic surveys and surface seismic and magneto-telluric surveys for a better understanding of local subsurface structure, and
- LiDAR coverage of the area to get a bare-earth surveys and decipher surface lineaments/faults as well as drainage patterns. It was further suggested that after these investigations have been carried out, the new knowledge gained would be helpful in developing the plan of setting up a bore-hole laboratory.

The ICDP agreed to extend help to the project with their technical expertise in deep drilling and logging, training of manpower and support towards drilling. The support from MoES to the program was overwhelming. During 2011–2014 the following was accomplished:

- Nine exploratory boreholes were drilled through the entire basalt column and penetrating to 300–500 m into the granitic basement.
- The entire reservoir triggered seismicity zone and nearby area were covered by detailed magneto-telluric surveys for estimating the thickness of the basalt cover and its lateral variations.
- For the first time, airborne gravity gradient surveys were carried out over the Koyna region. Additionally, airborne magnetic surveys were also conducted for a better under- standing of the subsurface structure
- The region was covered by LiDAR surveys which provided a bare-earth visualization, and mapping of fault-related geomorphology
- Out of the 9 drilled boreholes, 6 were instrumented with matched 3-component seismometers for improving hypocentral parameters. All the seismometers were place at around 1500 m depth, well below the basalt cover
- Detailed heat flow measurements were carried out in the boreholes to have a better estimate of subsurface temperatures.

All these investigations were extremely helpful in our understanding the subsurface structure as well as improving the location of earthquakes to $\pm$ 300 m accuracy. Major results of these 3-year-long investigations were published in Bulletin of the Seismological Society of America. Some of the salient results are mentioned.

(a) Discovery of near absence of the Mesozoic sediments below the basalt cover that had been long debated.

(b) Basalt cover thickness in the region is related to topography. While the surface topography has variations of – 1000 m, the basement is found to be flat at present lying – 300 m below the current sea level.
(c) The temperatures at a depth of 6 km were estimated to be – 130 °C, indicating a rather cold and brittle crust.
(d) Earthquakes' hypocenters concentrate around subsurface sharp density and resistivity contrast.
(e) MT surveys delineated mechanically weak subsurface zones as well as tracing the basalt–basement contact.
(f) Alignment of basement faults was revealed through the detailed geological/geophysical surveys.
(g) LiDAR surveys helped in generating bare-earth model of the region as well as delineating the Donachiwada fault that hosted the 1967 M 6.3 and several M ~ 5 earthquakes.

During 16–18 May 2014 the second ICDP workshop was held at Koyna as a sequel to the first workshop in which the work done by the Indian scientists was reviewed and a broad framework of the activities for the next phase leading to deep drilling was worked out. The Workshop was well attended by 49 experts that included 12 scientists from abroad representing 8 countries. The scientific questions discussed, several of them having been taken from the San Andreas Fault Observatory at Depth (SAFOD) program [Zoback M., Hickmann S. and Ellsworth W., SAFOD's Science Team (2011)]. Scientific drilling into the San Andreas Fault zone: an overview of SAFODS's first five years (Sci. Drill, 11, 14–28) included:

1. What is the fluid pressure and permeability within and adjacent to the fault zone?
2. What are the composition and origin of fault-zone fluids and gases?
3. How do stress orientation and magnitude vary across fault zones?
4. How do earthquakes nucleate?
5. How do earthquake ruptures propagate?
6. How do earthquake source parameters scale with magnitude and depth?
7. What is the role of water reservoirs in triggering earthquakes?
8. What is the 3-D/4-D nature of the fault zone?

The most important first step was taken by the Ministry of Earth Sciences was to create an institute of Borehole Geophysics Research Laboratory (BGRL) at Karad, near Koyna in 2014 with a mandate to build necessary R&D infrastructure and expertise to carry out scientific drilling investigations, including planning and execution of the Koyna deep drilling program. It was agreed that in the first phase a 3 km deep Pilot Borehole should be drilled. The results of the investigations and the measurements conducted in and around this Pilot Borehole would provide the necessary inputs for design the final 6 to 7 km deep borehole in the earthquake zone. The location of the Pilot Borehole was chosen keeping in mind the repeating of $M \geq 2$ target earthquakes, focal depths (preferably > 7 km), easy accessibility for all kind of measurements.

The BGRL took the responsibility of drilling the 3 km deep Pilot Borehole. It was started on 20 December 2016 and completed on 11 June 2017. As was the case with other boreholes in the region, no sediments were found below the Deccan Trap basalt column in the Pilot Borehole, like the other boreholes drilled in the region. Zones of immense water inflow were encountered while drilling. An online gas analysis facility was set up by with the help of ICDP. Geophysical logging was conducted, and cores were obtained from several depths for detailed studies. These included 9-m-long 4-inch-diameter cores from the depths of 1679, 1892 and 2091 m. There was almost 100% core recovery as reported in several publications.

5.1.1 ICDP-Post-Operation Workshop

A Post-Operational ICDP workshop was held at Pune, India during 14 to 16 October 2017. It was attended by 67 participants including 19 from Canada, France, Germany, Italy, Japan, the Czech Republic, South Korea, Spain, Switzer- land, and the USA. There was a total satisfaction with the results obtained and plans of further work leading to designing the proposed 7 km deep Main Borehole. Participants of the workshop visited Borehole Geophysics Research Laboratory at Karad. The Core Repository was of immense interest. On the final day, detailed recommendations for the future course of action were made focusing on developing of Koyna RTS information system; characterization of Donachiwada fault; deployment of a vertical string of seismometers below 1500 m in the Pilot Borehole; a variety of hydrogeo-logical/geophysical/geological measurements to be made; detailed study of the core samples; setting up of a reference borehole outside Koyna RTS area; and preparation of a detailed plan of the proposed deep bore laboratory at a depth of ~ 7 km. Overall, the placing of the borehole at a very appropriate location and carrying out all the necessary measurements during the drilling of 3-km-deep borehole were appreciated.

The work was going on in full swing after the third ICDP workshop when the Covid-19 pandemic put brakes on the progress and broke the momentum. During the last 3 years of uncertainty that is continuing has resulted in certain setbacks in meeting the planned targets. Some of the planned targets and the timelines were worked out keeping in view the conditions prevailing in the country and in the world. I am hopeful, that a dream of drilling a 7 km deep borehole right into the heart of the seismic zone to catch a *Live* earthquake may become a reality someday (Fig. 5.1).

Photo 5.1ICDP Post-Operation International Workshop on Scientific Deep Drilling in Koyna, India

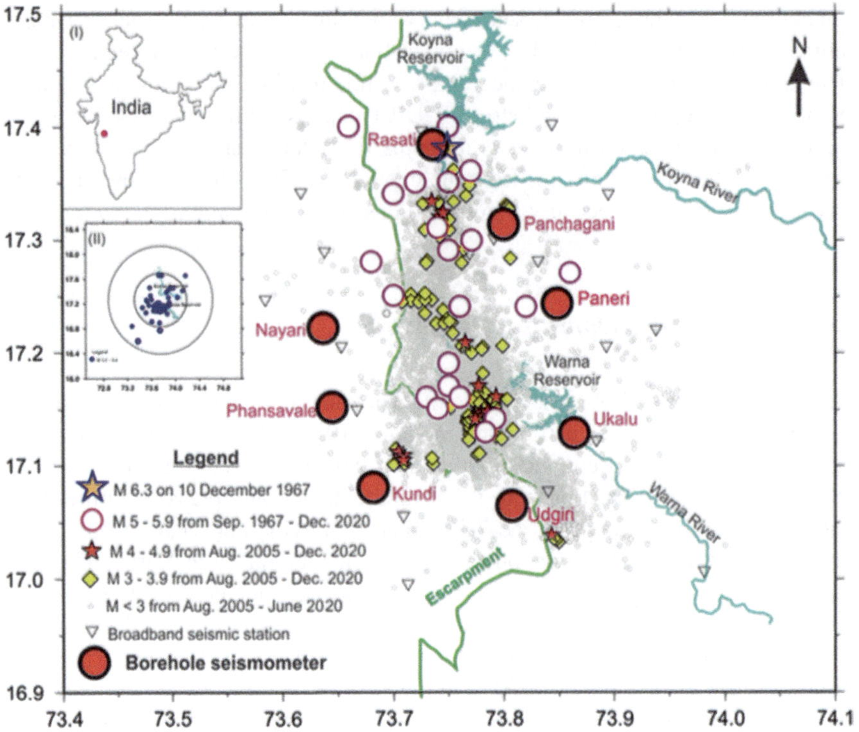

Fig. 5.1 Earthquakes in the Koyna region, India; surface and borehole seismic stations (there are 17 surface and eight borehole stations currently operating); the Western Ghat Escarpment on the Donachiwadi fault (updated from Gupta 2017). Inset I: Location of Koyna in India. Inset II: $M \geq 3.7$ earthquakes for the period 1967–2020 in the vicinity of Koyna (USGS), and an outer circle of 100 km indicating almost no activity outside the Koyna region. The pilot borehole is located in the close proximity of the Donachiwadi fault (Fig. 5.2; Photo 5.1)

5.2 Developing an Earthquake Resilient Society[1]

It is a dream of every seismologist that one-day earthquake prediction may become a reality and precious human lives are saved. The failed experiments and hopes raised by initial precursory studies in the seventies, with no remarkable success, have reduced the enthusiasm of the scientists, world over, to the possibility of making short-term predictions. It is amply clear that the mechanism relating geophysical and geochemical precursors or even anomalous animal behaviour prior to earthquakes is highly non-linear and requires a perfect understanding of a complex system, which has several unknowns. This realization has convinced the scientific community that earthquake prediction is a very distant goal or may be it is a mirage, which we are chasing, only time will tell. There is a need to change our approach from a prediction

[1] Based on: Harsh K. Gupta and Kanchan A. Sabnis (2021), 'Developing an Earthquake Resilient Society in the Vicinity of Himalaya', J. Geological Society of India, 97, pp 1593–1602

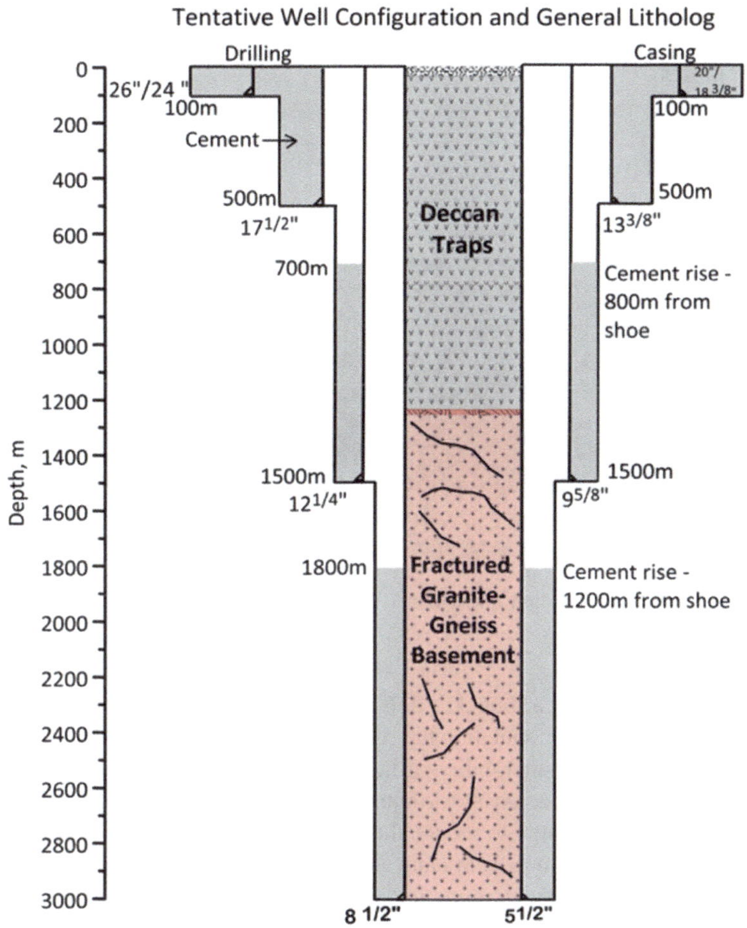

Fig. 5.2 Well Configuration for the pilot boreholes

base to a response base with the primary aim to save lives and minimize financial losses.

This realization has made the scientists turn their attention to earthquake hazard assessment and mitigation that necessitate joining hands with structural and civil engineers to design safe houses for people. The positive results of this approach is clearly visible in advanced countries like Japan, USA and New Zealand where hardly any person dies due to moderate earthquakes of magnitude 6 or so, as their houses are designed to withstand the earthquake forces. In India too, this approach is being adopted, especially after the 2001 Bhuj earthquake, by studying strong ground motions during earthquakes and their impact on the existing and newly planned structures in the country. The process needs to be speeded up.

Photo 5.1 ICDP Post-Operation International Workshop on Scientific Deep Drilling in Koyna, India

The old saying that 'earthquakes don't kill, but buildings do' continues to hold good. The ever-growing population has resulted in an increase in built environment, further increasing the vulnerability of the people to earthquake hazard. Thus, the requirement of building a robust earthquake resilient society comes to the fore. In the recent times an example to reduce the vulnerability of people to earthquakes in developing countries comes from Haiti in the Caribbean region. In 2010 an earthquake of magnitude 7.0 killed more than ~ 300,000 people and a part of the country was totally devastated. The authorities, on a war footing, took the earthquake countermeasures after the earthquake. As a result, when Haiti was jolted once again by a larger 7.2 magnitude earthquake in 2021near the same location of 2010 earthquake, loss of human lives was ~ 2200 deaths, a miniscule in comparison to a slightly smaller magnitude earthquake just 10 years before.

I was appointed as a Member, National Disaster Management Authority of India (NDMA) in 2011. At that time, it appeared very worthwhile to initiate a project with a sound scientific study to develop a plausible Earthquake Scenario for the earthquakes occurring in the Himalayan region of India. The objective was to estimate the likely loss of human lives if a great earthquake occurs in the earlier locations in the Himalayan earthquake belt. This information was planned to be shared with all stake holders for awareness and do the needful to enhance the earthquake resilience in the areas likely to be affected. The project was to be executed through a public private partnership towards building earthquake resilient society in the vicinity of the Himalayan earthquake belt. The reason being that during a short interval of 53 years, from 1897 through 1950, four damaging earthquakes of magnitude ~ 8 occurred in the

Himalayan earthquake belt, namely the 1897 Shillong earthquake, the 1905 Kangra earthquake, the 1934 Bihar-Nepal earthquake and the 1950 Assam earthquake.

Towards this goal, NDMA undertook two sequential projects. The main component of the project included generating a scientific scenario based on a major earthquake from history, assess the damage based on the built environment, population density, train various stakeholder on mitigation and preparedness measures and finally culminate in a mega-mock drill covering major cities of the Himalaya region. Phase One covered the Northwestern states of Himachal Pradesh, Punjab, Haryana, and UT Chandigarh for the repeat of the 1905 Kangra earthquake. This magnitude ~ 8 earthquake had done widespread damage from Kangra to Dehradun and claimed over 20,000 human lives. In Phase Two the northeastern states of Assam, Meghalaya, Manipur, Mizoram, Tripura, Arunachal Pradesh, Sikkim, and Nagaland were covered for the repeat of the 1897 Shillong earthquake. The 1897 Shillong earthquake of magnitude 8.7 has been one of the strongest continental earthquakes having generated visible surface waves, first time reported globally, and seiches (oscillations in water level) in lakes all over Europe.

For the first phase, a scenario was developed as to what would happen if the 1905 magnitude ~ 8 Kangra earthquake repeats today. An earthquake scenario is a planning tool that helps to understand the anticipated losses due to an earthquake and aids in planning how to minimize its impact on life, property, and economic activity in the region. A successful scenario tells the story of a defined earthquake and its specific impacts. It helps decision makers and the public visualize specific impacts that are based on sound scientific and engineering knowledge. Based on scenario projections, the mitigation and response plans are prepared and evaluated. The tool provides an opportunity to examine alternative options and stimulate creative thinking about the need for new policies and programs. Incorporating the latest scientific, engineering, and societal knowledge about a region's seismic hazard based on anticipated acceleration and displacement during an earthquake and considering the regional characteristics (such as, soil characteristics, building types, health of lifeline buildings, and population characteristics), a scenario creates a compelling picture that members of the local community can recognize and relate. An earthquake scenario thus improves awareness about an earthquake and its likely impact on a community and the region.

In consultation with scientists and engineers, a scenario for the repeat of 1905 Kangra earthquake was developed. The epicenter of the hypothetical M 8 earthquake was placed at Mandi (very close to the epicenter of 1905 Kangra earthquake), the focal depth was fixed at 15 km, which would cause strong shaking in whole of Himachal Pradesh and UT Chandigarh as well as most parts of Punjab, Haryana, and in some parts of Jammu and Kashmir, Uttarakhand, and Uttar Pradesh. However, for administrative and logistic constraints, the exercise was restricted to Haryana, Himachal Pradesh, Punjab, and UT of Chandigarh. The maximum earthquake intensity was estimated to be ~ X on the MSK scale. The table below gives the estimated number of human lives lost and injured in the states of Haryana, Himachal Pradesh, Punjab, and the Union Territory of Chandigarh (Table 5.1).

Table 5.1 The anticipated number of injuries and deaths due to Mandi Mw 8 scenario earthquake

State/UT	Population (Census 2011)	Injured (approximate)	Deaths (approximate.)
Haryana	25,351,462	1,489,800	311,700
Himachal Pradesh	6,864,602	1,108,600	159,000
Punjab	27,743,338	2,484,600	464,100
Chandigarh	1,055,450	126,700	24,300
Total	61,014,852	5,209,700	959,100

I was quite worried to see the high number of casualties totaling close to a million, in case the earthquake occurred in the middle of the night. However, the Muzaffarabad earthquake of M 7.6, that occurred on 8 October 2005 claimed ~ 87,000 human lives. The earthquake had occurred near India–India border in Pakistan. It had occurred in the morning at 08.50 local time, when people are awake and mostly away from their homes. Had it occurred in the middle of the night, the number of the human lives lost would have been much higher. Also, the magnitude was 7.6. The energy released by a magnitude 8 earthquake is almost 10 times more than a magnitude 7.6 earthquake. These two issues made me believe that the lives likely to be lost if a M 8 earthquake occurs in the middle of the night in Mandi is not erroneous.

The first project was launched by the Prime Minister of India, Dr. Manmohan Singh in June 2012. The Chief Ministers of Punjab, Haryana, Himachal Pradesh and the Chief Administrator, UT Chandigarh along with their state machinery extended full support to the project. One representative was attached with each state and UT for liaison and data collection. The Project included extensive Training of Engineers to understand their built environment; Capacity Development Program for various administrative officials, local population, specifically school students and children; Awareness Campaign involving local and national celebrities through print, audio, and video mode to reach the masses; and finally, the entire exercise was to culminate in the conduct of the multi-state mega mock exercise.

During the preparatory phase of Mega Mock exercise, representatives from all 45 districts of the participating States and UT Chandigarh underwent the training organized at Chandigarh and Shimla. In the run-up to this exercise, the NDMA, in collaboration with the participating States and UTs launched an extensive public awareness campaign through print media, electronic media, social media, outdoor advertising, radio advertising, and SMS alerts. Hoardings carrying safety messages were displayed at prominent bus stands, public gardens, malls, road crossings and near the government offices. Young volunteers joined hands with the administration in spreading awareness. Nukkad Natak (a short drama play at the corner of a street), rallies by school children and safety carnivals were also organized. Radio jingles conveying earthquake safety tips were aired. Safety messages and information about the campaign by NDMA Members were also repeatedly broadcasted. The system of bulk SMS, which makes it possible to reach a large group of contacts within seconds, was also tapped to ensure maximum awareness. Celebrities like Mr. Yuvraj

Singh, Mr. Abhinav Bindra, and Subedar Major Vijay Kumar, AVSM, SM joined the campaign as goodwill ambassadors to attract more publicity for the event.

The first scientifically steered, Multi-State Mega Mock Exercise in the country was conducted on 13 Feb 2013 with the hypothetical M 8 earthquake occurring at 11.30 am and observations being made in the tri-cities of Chandigarh, Panchkula (Haryana) & Mohali (Punjab), and Shimla (Himachal Pradesh). The scenarios generated under the Project formed the basis for this exercise. NDMA Members and experts, Special Secretary DDMA, Professors from IITs, DG and senior officers from NDRF, Adviser to the Governor, UT Chandigarh, FCRs of Haryana and Punjab, ACS of HP, and their teams of DCs, SDMs and ADMs, Fire Service, Medical Service, Police Forces, Transport department, State Electricity Board, Telecommunication department, Civil Defense Volunteers, Army, and Home guards took part in the exercise. The drills were set up to emulate real-life earthquake conditions. 60 locations including hospitals, government offices and buildings, malls, schools, engineering college, office complex, airports, railway stations, busy market areas, bus stations, cinema halls and petrol stations were reported as affected with the hypothetical earthquake.

To this call of distress, district, and state first responders started responding in phases. State and district emergency operations centers (EOCs) and ESFs were activated, and top officials from several departments activated their respective action plans. As per the scenario, normal communication lines were badly affected and state EOCs, city EOCs, and the unified command used wireless communication for information sharing. Later, the central forces aided the state authorities as well. For an immediate response, resources from numerous government and private institutions were mobilized. NDRF teams were deployed to several places in response to a search and rescue request from the State EOC, where they showed lifesaving and search and rescue skills. The Observers from the Army, comprising officers, JCOs and ORs were deputed across all the cities to carry out an independent third-party audit. They monitored and observed the entire event. Later, during Hot wash-up, the observers highlighted the positive points of the entire exercise as well as the areas which required improvement and better coordination was needed. Press briefings were organized to provide comprehensive coverage.

Similar Multi-State Mock Exercise was repeated in the northeast India for the repeat of 1897 Shillong earthquake. This earthquake in 1897 had claimed only 1500 human lives. The reason being a very low population density in the northeast India region. The isoseismals prepared for this earthquake are presented and the table provides the estimated of the anticipated human lives lost if the earthquake occurred in the middle of the night. The procedures followed for the Kangra earthquake mock drill were repeated for the 8 states in the northeast Indian region. However, in view of large extent of the region, varying terrain, and constraints of transportation and logistics, the Mock Exercise was conducted in two phases covering 17 cities. In Phase I on 10 March 2014, states of Assam (Guwahati and Jorhat), Arunachal Pradesh (Itanagar and Ziro), Meghalaya (Shillong and Tura) and Sikkim (Gangtok and Mangan.) were covered. In Phase II on 13 March 2014, Nagaland (Kohima and Dimapur), Manipur (Imphal and Thoubal), Tripura (Agartala and Ambassa), Mizoram (Aizawal and

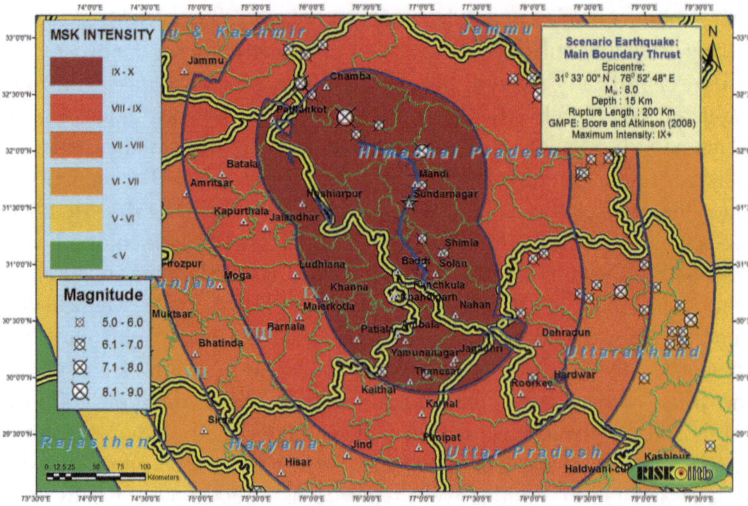

Fig. 5.3 Intensity distribution of a scenario Mw 8 earthquake depicted in MSK Intensity Scale, with the rupture centered at Mandi, Himachal Pradesh, near the epicenter of 1905 Kangra earthquake. Epicenters of the recent earthquakes occurring in the region are shown

Serchip) and Assam (Silchar) were covered. All cities had pre-identified locations for the drill and the community was also made aware of the exercise through awareness generation campaign. This exercise was done to observe the inter and intra agency coordination and the overall mobilization of resources and their optimum utilization. The Mock Exercise was conducted by NDMA jointly in participation with State administration, NDMA Members and experts, CSIR-NEIST Jorhat, DG and senior officers from NDRF, Senior officers and their teams from NE States, NEC, NEEPCO, NEHU, NDRF, Defense services, ITBP, BSF, CRPF, Assam Rifles, Indian Red Cross, paramedics, and NGOs. This exercise broadly covered 80 plus locations which included Hospitals, government offices/buildings, malls, schools, engineering college, office complex, airport, railway station, busy market areas, bus station, cinema halls and petrol stations.

These exercises have been a landmark in NDMA's efforts in developing an earthquake resilient society in the vicinity of Himalayan seismic belt. It has been recommended that similar exercises need to be conducted in other parts of the country, where disastrous earthquakes have occurred in the past. It is also important to observe an earthquake awareness day, for example on 5 April of every year in the north-west Indian states as the Kangra earthquake occurred on 5 April 1905, and 12 June every year for the north-east Indian states as the Great Shillong earthquake occurred on 12 June 1897. There is a need to target the school children in these exercises so that the future society would be more knowledgeable about earthquakes and learn to live safely with them (Figs. 5.3 and 5.4; Photo 5.2; Table 5.2).

DAILY Chandigarh › THURSDAY 14 FEBRUARY 2013 | 03

"There were four injured persons whom we referred to GMSH-16 and discharged them in a mock drill."

DR HARPANT SINGH, Surgeon, GMSH, Sector 16

regional

Mega Mock Drill exposes chinks in disaster management plan

DP CORRESPONDENT
Chandigarh

The Mega Mock Drill on earthquake preparedness, conducted in tri-cities of Chandigarh, Mohali and Panchkula on Wednesday, have brought out many loopholes in the preparation of the various departments of the Administration. The exercise was held at 15 different places of the city.

In sector 17 the ambulances reached late to its schedule time and on another incident people got confused while coming out from the buildings in the absence of the exit or entry signs. There was no public address system to inform the public at sector 17. The move to vacate the parking was also wrong as the exercise should have done in the natural environment.

The lack of the women force in the civil defence system was also felt. The lack of the coordination was also noticed in the UT secretariat as the team of the NDRF was unable to locate the incident commander. The lack of the public awareness was also felt as instead of participating in the exercise many went for the market for the shopping.

The lack of establishment Disaster Management Plans at various levels was also oberved. The exercise was conducted with collaboration and guidance of National Disaster Management Authority (NDMA).

The officers of the NDMA said that the key lessons learnt are : Need for state-of-the art emergency operation centres at the state and district levels with redundancy; Setting-up and use of wireless communication system for the stakeholders to more effectively deal with an earthquake disaster scenario; Recognise capacity of having better road accessibility in case of disaster; Identification of requirements of state-of-the-art equipment and systematic inventory of resources, both with government and private sector; Need for adequate number of ambulances and greater public awareness and sensitisation of people for enhanced preparedness.

The need to make younger generation aware of the earthquake hazards and coping mechanism is also felt. About 900 personnel from NDRF organised in 30 teams, 900 civil defence personnel and 270 independent observers from Army participated in the exercise. The emergency support functionaries like police, health, fire, ambulance, civil defence, power and PWD etc also participated.

Mega Mock Drill at Sector-17 plaza in Chandigarh on Wednesday. UPENDRA SEN GUPTA

Exercise held across Panchkula

DP CORRESPONDENT
Panchkula

District Administration in active collaboration with NDRF conducted a mega Mock Drill at 15 various sites in Panchkula on Wednesday. Around 1500 officers and personnel of various departments participated in the drill which was monitored and supervised by Harsh Gupta, member NDMA, P Raghvendra Rao, IAS, Principal Secretary Urban Local Bodies, Haryana, Ashima Brar, Deputy Commissioner, and some of the Army and Air Force observers.

The teams demonstrated host of activities like creating suspension bridges with ropes and harnesses and taking out uninjured victims by helping them slither. They also showed how a less injured could be brought down from the affected building by loading stretchers on these bridges. For those who remained trapped in little space under the debris, the Jawans showed how the victim's hands and feet could simply be fastened around the body to bring down the victim. On the occasion, Deputy Commissioner informed that district administration had made extensive arrangements for the successful demonstration and completion of such mock drills. She told that basic scenario of such mock drill was earthquake and outbreak of fire. Services of around 20 fire fighting systems, 19 Ambulance, 15 JCB machine and various water tankers along with staff was utilised for the successful completion of these drills.

Team of civil defence bringing the injured who was trapped on a building on mock drill conducted by district administration in Panchkula. DP PHOTO

Photo 5.2 There was a very wide media coverage following the mega-mock drill on 13 February 2013

Fig.5.4 Intensity distribution for the MSK intensity scale for the occurrence of 1897 Shillong earthquake

Table 5.2 The estimated number of the human lives lost and humans injured if the 1897 Shillong earthquake repeats in mid-night

S No	State	Injuries	Deaths
1	Assam	~ 40,00,000	~ 6,00,000
2	Manipur	~ 2,10,000	~ 38,000
3	Meghalaya	~ 4,80,000	~ 68,000
4	Nagaland	~ 1,80,000	~ 30,000
5	Sikkim	~ 84,000	~ 12,000
6	Tripura	~ 4,32,000	~ 66,000
7	Arunachal Pradesh	~ 76,000	~ 12,000
8	Mizoram	~ 1,10,000	~ 17,000
Total		~ 55,72,000	~ 8,43,000

A Few Concluding Words

Over the last decade, I have realized that despite all the technological, scientific, and administrative developments, human lives lost due to earthquakes and tsunamis continue to rise globally. But if we look back at the geological records one can see that while these natural hazards may not have increased but natural disasters, have. We have just completed the first 23 years of the 21st Century and both the human and financial losses have far exceeded that of the 20th Century and shows no signs of decline in spite of our best efforts and rather show an upward trend. Some reasons are obvious and some are beyond our control. My focus all these years has been to address the issues that are in our control and can minimize the impact of these hazards with our scientific and social interventions, if not completely avoided. I continue my efforts in this direction from the confines of my office at the National Geophysical Research Institute, Hyderabad, where I have spent almost one-half of my working years. Being an utterly optimistic person I am sure a dent would be made to address this life threatening problem (Photo 1).

Photo 1 CSIR-National Geophysical Research Institute, Hyderabad, India.

H. K. Gupta, *Harsh K. Gupta*, Springer Biographies,
https://doi.org/10.1007/978-3-031-63561-8

A Complete List of Publications

Research Publications

1. Gupta HK (1964) Direction of approach of short period microseisms at Shillong. Ind J Met Geophys 15:653–656
2. Gupta HK, Narain H (1967) Crustal structure in the Himalayan and Tibet Plateau region from surface wave dispersion. Bull Seism Soc Am 57:235–248
3. Gupta HK (1968) Lateral surface wave velocity variation studies, Proc 'Symp on Geophysics.' Indian Geophys Union 1968:139–152
4. Gupta HK, Sato Y (1968) Regional characteristics of Love wave group velocity dispersion in Eurasia. Bull Eq Res Inst Tokyo 46:41–52
5. Gupta HK, Mohan I (1968) The N.G.R.I. seismological observatory. Bull N.G.R.I. 6:125–133
6. Narain H, Gupta HK (1968) Observations on Koyna earthquake. J Ind Geophy Un 5:30–34
7. Narain H, Gupta HK (1968) The Koyna earthquake. Nature 217:1138–1139
8. Narain H, Gupta HK (1968) The how and why of Koyna. Sci Today 47–51
9. Gupta HK (1969) Studies on seismic wave dispersion, crustal structure of Himalayan region and global seismicity. Ph.D. Thesis, Roorkee University
10. Gupta HK (1969) Recent development in lateral surface wave dispersion studies. A Review Proc Sym Use of Gauribidanur Seismic Array Data, B.A.R.C., pp 58–78
11. Gupta HK, Narain H, Rastogi BK, Mohan I (1969) A study of the Koyna earthquake of Dec 10, 1967. Bull Seism Soc Am 59:1149–1162
12. Gupta HK (1970) Surface wave studies in India. Ind Geophy Union, Tandon Volume, pp 69–81
13. Gupta HK, Negi JG (1970) High magnitude global seismicity and lateral velocity gradients. Pure App Geophys 80:92–101
14. Gupta HK, Negi JG (1970) Lateral variation of Raleigh wave dispersion characteristics in Australia. Bull Seism Soc Am 60:1897–1906
15. Gupta HK, Mohan I, Narain H (1970) The Godavari Valley earthquake sequence of April 1969. Bull Seism Soc Am 60:601–615
16. Gupta HK, Sitaram MVD, Narain H (1970) Station factor of N.G.R.I. Observatory for teleseismic events. Bull N.G.R.I. 8(1&2):1–9
17. Gupta HK, Mohan I, Rastogi BK, Sitaram MVD, Narain H (1970) Seismological investigations under upper mantle project. Bull N.G.R.I. 8(3&4):1–18
18. Gupta HK, Rastogi BK, Narain H (1971) The Koyna earthquake of December 10, 1967: A multiple seismic event. Bull Seism Soc America 61:167–176
19. Mithal RS, Gupta HK (1971) Seismic zoning in India. Bull N.G.R.I. 9(1&2):9–22

H. K. Gupta, *Harsh K. Gupta*, Springer Biographies,
https://doi.org/10.1007/978-3-031-63561-8

20. Gupta HK, Rastogi BK (1972) Earthquake m_b versus M_s relations and source multiplicity. Geophysics J Roy Astr Soc 28:65–89
21. Gupta HK, Mohan I, Narain H (1972) The Broach earthquake of March 23, 1970. Bull Seism Soc Am 62:47–61
22. Gupta HK, Rastogi BK, Narain H (1972) Common features of the reservoir associated seismic activities. Bull Seism Soc Am 62:481–492
23. Gupta HK, Rastogi BK, Narain H (1972) Some discriminatory characteristics of earthquakes near the Kariba, Kremasta and Koyna artificial lakes. Bull Seism Soc Am 62:493–507
24. Gupta HK, Sitaram MVD, Narain H (1972) Surface wave and body wave magnitudes of some Sino Soviet nuclear explosions and earthquakes. Bull Seism Soc Am 62:509–517
25. Gupta HK (1973) Plate tectonics and lateral velocity gradients. Tectonophysics 16:145–154
26. Gupta HK (1973) Geophysical investigations for the Himalayan region undertaken at the N.G.R.I.. In: Proceedings seminar on geodynamics of Himalayan Region, geodynamic series, vol 5, pp 72–96
27. Gupta HK, Santo T (1973) Worldwide investigation of the mantle Raleigh wave group velocities, Part I, Dispersion data for new 31 great circle paths. Bull Seism Soc America 63:271–282
28. Gupta HK, Rastogi BK, Narain H (1973) A study of earthquakes in the Koyna region and common features of the reservoir associated seismicity, American Geophy. Union, Monograph 17:455–467
29. Gupta HK, Skoko D, Sato Y (1973) Accuracy of determination of epicenter and origin time of small magnitude earthquakes in the Indian sub-continent. Bull Seism Soc. Am 63:1901–1912
30. Gupta HK (1974) Some seismological observations and tectonics from Hindu Kush to Burma region. In: Proceedings 4th seminar on Himalayan geology. Wadia Institute of Himalayan Geology, India, vol 4, pp 465–480
31. Gupta HK, Rastogi BK (1974) Will another damaging earthquake occur in Koyna? Nature 248:215–216
32. Gupta HK, Rastogi BK (1974) Investigation of the behaviour of reservoir associated earthquakes, Colloquium on seismic effects of the reservoir impounding, The Royal Society, London. Eng Geol 8(1/2):29–39
33. Gupta HK, Hamada K (1975) Rayleigh and Love wave dispersion up to 140 sec period range in the Indonesia Philippine region. Bull Seism Soc Am 65:507–521
34. Singh DD, Rastogi BK, Gupta HK (1975) Surface wave radiation pattern and source parameters of Koyna earthquake of December 10, 1967. Bull Seism Soc Am 65:711–731
35. Gupta HK (1976) Seismological investigations and tectonics of the Kashmir Hindu Kush Pamir region. In: Invited paper presented at the international colloquium on geotectonic of Kashmir, Karakorum, Hindu Kush, Pamir Orogenic belts, Proceedings published by Academic Nazionale Dei Lincei, Rome, vol 21, pp 42–66
36. Gupta HK, Combs J (1976) Continued seismic activity at the Koyna reservoir site, India. Eng Geol 10:307–313
37. Gupta HK, Nyman DC (1976) Long period Rayleigh wave group velocities along a great circle path in Southern Asia. Geoviews 2:15–27
38. Gupta HK, Nyman DC, Landisman M (1976) Raleigh wave group velocities between New Delhi, India and Shiraz, Iran, extending to long periods. In: Drake CL (ed) Geodynamics: progress and prospects, American Geophysical Union, pp 121–127
39. Gupta HK (1977) Earthquake prediction: facts and fantasies. J Indian Geophys Union XIV(1–4):115–125
40. Gupta HK, Nyman DC (1977) Short period surface wave dispersion studies in the East Mesa Geothermal field. California, Geotherm Resour Council, Trans 1:123–125
41. Gupta HK, Nyman DC, Landisman M (1977) Shield like upper mantle velocity structure below the Indo Gangetic Plains: inferences drawn from long period surface wave dispersion studies. Earth Planet Sci Lett 34:51–55
42. Gupta HK, Nyman DC, Landisman M (1977) Shield like upper mantle structure inferred from long period Rayleigh and love wave dispersion investigations in the Middle East and South East Asia. Bull Seism Soc Am 67:103–119

43. Nyman DC, Gupta HK, Landisman M (1977) The relationship between group velocity and phase velocity for finite, discreate observations. Bull Seism Soc Am 67:1249–1258
44. Gupta HK (1978) Recent long period surface wave dispersion investigations in the Himalaya and Neighboring Regions, Indian National Committee for International Union of Geodesy and Geophysics. J Ind Geophys Union 163–181
45. Gupta HK, Combs J (1978) Investigation of isoseismals for some large magnitude earthquakes in China. Bull Seism Soc Am 68:193–204
46. Singh DD, Gupta HK (1979) Source mechanism of two Burma India border earthquakes. Bull Ind Soc Eq Technol 16:63–74
47. Singh DD, Rastogi BK, Gupta HK (1979) Spectral analysis of body waves for earthquakes and their source parameters in the Himalaya and nearby regions. Phys Earth Planet Intr 10:143–152
48. Singh DD, Gupta HK (1979) Source mechanism and surface wave attenuation studies for Tibet earthquakes of July 14, 1973. Bull Seism Soc Am 69:737–750
49. Gupta HK, Singh VP, Mohan I (1979) Earthquake prediction research in Assam, Northeast India. In: International symposium on Eq. Prediction, Paris, 1979 SC 79/Conf. 801/Col.14/I 9, pp 1–15
50. Gupta HK, Rastogi BK, Ramakrishna Rao CV (1979) Reservoir induced seismicity. Geophys Res Bull 17:391–404
51. Gupta HK, Singh DD, Divakara Rao V, Chalam SV (1979) Seismological, geochemical and gravity investigations in the Himalaya and nearby regions. Geophys Res Bull 17:337–360
52. Mohan I, Sitaram MVD, Gupta HK, Raju IMP, Prem Kishore D, Suryaprakasam K, Rao CVR, Ramachandra Rao SVR, Khamruddin MC (1979) Operation of seismological observatories and some related studies. Geophy Res Bull 17:375–383
53. Gupta HK, Singh DD (1980) Spectral analysis of body wave for earthquakes in Nepal Himalaya and vicinity: their focal parameters and tectonic implications. Tectonophysics 62:53–66
54. Gupta HK, Singh VP (1980) Teleseismic P wave residual investigations at Shillong, India. Tectonophysics 66:T19–T27
55. Gupta HK, Singh DD (1980) Source mechanism study of Quetta earthquake of May 30, 1935. Geol Bull Univ Peshawar 13:143–150
56. Gupta HK, Rao CVR, Rastogi BK, Bhatia SC (1980) An investigation of earthquakes in Koyna region, Maharashtra, for the period October 1973 through December 1976. Bull Seism Soc Am 70:1833–1847
57. Singh DD, Gupta HK (1980) Source dynamics of two great earthquakes of the Indian Subcontinent: The Bihar Nepal earthquake of January 15, 1934, and the Quetta earthquake of May 30, 1935. Bull Seism Soc Am 70:757–773
58. Singh DD, Gupta HK (1980) Source mechanism study of Burma border earthquake of October 17, 1969. Tectonophysics 67:139–152
59. Gupta HK, Bhatia SC (1981) A review of the long period surface wave studies in the Himalaya and nearby regions. Zagros Hindu Kush Himalaya Geodynamic Evolution, Geodynamics series, American Geophysical Union Geological Society of America Publication, vol 3, pp 294–306
60. Mohan I, Sitaram MVD, Gupta HK (1981) Some recent earthquakes in Peninsular India. J Geol Soc India 22:292–298
61. Gupta HK, Singh VP (1982) Is Shillong region, Northeast India, Undergoing a dilatancy stage precursory to a large earthquake? (Extended abstract). Tectonophysics 85:31–33
62. Gupta HK, Rastogi BK, Mohan I (1982) Studies on reservoir induced seismicity and earthquakes in Peninsular India. Geophy Res Bull 20:127–139
63. Gupta HK, Rao VD, Singh J (1982) Continental collision tectonics: evidence from the Himalaya and the neighboring regions. Tectonophysics 81:213–238
64. Gupta HK, Rastogi BK, Singh DD (1982) Seismological investigations in the Himalaya and nearby regions. Geophy Res Bull 20:141–155
65. Gupta HK, Ward RW, Lin TL (1982) Seismic wave velocity investigations at the Geysers Clear Lake Geothermal Field, California. Geophys 47:819–824

66. Singh DD, Gupta HK (1982) Q structure beneath the Tibetan Plateau from the inversion of love and Rayleigh waves attenuation data. Phys Earth Planet Inter 29:183–194
67. Gupta HK (1983) Induced seismicity hazard mitigation through water level manipulation at Koyna, India: a suggestion. Bull Seism Soc Am 73:679–682
68. Gupta HK, Iyer HM (1984) Are reservoir induced earthquakes of magnitude 5.0 at Koyna, India, preceded by pairs of earthquake of magnitude > 4.0? Bull Seism Soc Am 74:863–873
69. Gupta HK (1984) Seismicity in the vicinity of dams on Himalayan rivers and the problem of reservoir induced earthquakes. J Geol Soc India 25:85–93
70. Gupta HK (1984) The third Indian scientific expedition to Antarctica. J Geol Soc India 25:543–556
71. Gupta HK, Singh SC, Dutta TK, Saikia MM (1984) Recent investigations of North East India seismicity. In: Gu G, Ma X (eds) Proceedings: International symposium on continental seismicity and earthquake prediction. Seismological Press, Beijing, pp 63–71
72. Gupta HK (1984) Building of Dakshin Gangotri. Sci Today 18:17–23
73. Gupta HK (1985) Cachar earthquake of December 31, 1984 is it a signal for the beginning of seismic activity? J Geol Soc India 26:145–147
74. Gupta HK (1985) The present status of reservoir induced seismicity investigations with special emphasis on Koyna earthquakes. Tectonophysics 118(3&4):257–279
75. Gupta HK, Rajendran K (1986) Large artificial water reservoirs in the vicinity of the Himalayan foothills and reservoir induced seismicity. Bull Seism Soc Am 76(1):205–215
76. Gupta HK, Rajendran K, Singh HN (1986) Seismicity of the North East India Region. Part I: the data base. J Geol Soc India 28(5):345–365
77. Gupta HK, Singh HN (1986) Seismicity of the North East India Region, Part II: Earthquake Swarms precursory to moderate magnitude to great earthquakes. J Geol Soc India 28(5):367–406
78. Gupta HK, Bhatia SC (1986) Seismicity in the vicinity of the India Burma Border: evidence for a sinking lithosphere. J Geodyn 5:375–381
79. Nair PK, Gupta HK (1986) Dakshin Gangotri: The Indian permanent station in Antarctica. Third Indian expedition to Antarctica, Scientific Report 1986, Department of Ocean Development, Technical Publication No 3, pp 15–17
80. Gupta HK, Varma AK (1986) Magnetic characteristics in Antarctica over geological contacts in Schirmacher Hill Region and the ice shelf near Dakshin Gangotri (70° 05'37"S, 12° 00'00"E). In: Third Indian Expedition to Antarctica, Scientific Report, 1986. Department of Ocean Development, Technical Publication No 3, pp 231–239
81. Rajendran K, Gupta HK (1986) Was the earthquake sequence of August 1975 in the vicinity of Lake Oroville, California, reservoir induced? Phys Earth Planet Interior Amsterdam 44:142–148
82. Rastogi BK, Rao CVRK, Chadha RK, Gupta HK (1986) Microearthquakes near Osmanagar Reservoir, Hyderabad, India, Physics Earth Planet. Inter 44:134–141
83. Gupta HK (1987) The third Indian scientific expedition to Antarctica: some achievements. In: Contribution in marine sciences, Special Volume, NIO, Goa, pp 447–456
84. Gupta HK (1988) Medium term earthquake prediction. EOS 69(49):1620 and 1630
85. Gupta HK, Froidevaux C, Ansel V (1989) A zone of very low seismicity in the vicinity of Chang Thang platform. Tibet, Phys Earth Planet Inter 58:82–86
86. Gupta HK (1989) (with several other authors, listed in alphabetical order), Global patterns of intraplate stress: a status report on the world stress map project of the international lithosphere program. Nature 341:291–298
87. Rajendran K, Gupta HK (1989) Seismicity and tectonic stress field of a part of the Burma Andaman Nicobar Arc. Bull Seism Soc Am 79(4):989–1005
88. Gupta HK, Singh HN (1989) Earthquake Swarms precursory to moderate to great earthquakes in the northeast India region: Tectonophysics 167:285–298
89. Gupta HK, Fleitout L, Froidevaux C (1990) Lithospheric Subduction beneath the Arakan Yoma Fold Belt: quantitative estimates using Gavimetric and seismic data. J Geol Soc Ind 35(3):235–250

90. Gupta HK (1991) The August 8, 1988 earthquake of M = 7.3 in the North east India: a medium term earthquake prediction comes true. In: Proceedings, international seminar on earthquake prediction and hazard mitigation technology, science and technology agency, Japan, pp 407–426
91. Gupta HK (1992) Are RIS events of M > 5 preceded by a couple of foreshocks of M > 4? Bull Seism Soc Am 82(1):517–520
92. Gupta HK (1992) Earthquake prediction related studies in India. In: Special Issue on seismology in India-An overview. current science, vol 62, no 1&2, pp 257–263
93. Gupta HK (1992) Reservoir induced Earthquakes. In: Special issue on seismology in India - An overview. current science, vol 62, no 1&2, pp 183–198
94. Rajendran K, Talwani P, Gupta HK (1992) State of stress in the Indian subcontinent. A review. In: Special issue on seismology in India an overview current science, vol 62, no 1&2, pp 86–93
95. Gupta HK (1992) The Malawi earthquake of March 10, 1989: a report of the macroseismic survey. Tectonophysics 209:165–166
96. Gupta HK (1993) Earthquake hazard assessment in the Himalayan Region. In: Qasim SZ (ed) Science and quality of Life. The Offsetters, New Delhi, pp 529–544
97. Gupta HK (1993) Patterns preceding major earthquakes in north east India (Spl Issue). Curr Sci 64(11&12):889–893
98. Gupta HK (1993) The Deadly Latur earthquake. Science 262:1666–1667
99. Gupta HK (1993) Seismic hazard assessment in the Alpide Belt from Iran to Burma. Ann Geofis XXXVI(3–4):61–82
100. Gupta HK, Mohan I, Rastogi BK, Rao MN, Ramakrishna Rao CV (1993) A quick look at the Latur earthquake of 30 September 1993. Curr Sci 65(7):517–520
101. Gupta HK, Verma SK (1994) India's contribution to geophysical investigations in Antarctica. Memoir Geol Soc India (Spl Issue)
102. Gupta HK (1994) Artificial water reservoirs and earthquakes. In: Proceeding of 81st Indian Science Congress Part II: Presidential Address, Section VII: Earth Systems Sciences, pp 1–16
103. Gupta HK (1994) Where the Indian and Eurasian plates continue to collide. In: Proceeding of WSSI workshop, seismic risk management for countries of the Asia Pacific Region, Bangkok, Thailand, Incede Report, pp 85–101
104. Gupta HK (1994) A note on 'Stable Continental Region Earthquakes.' J Geol Soc India 43:619–620
105. Gupta HK (1995) Stable continent region earthquakes with special emphasis on Latur Earthquake of September 30, 1993. In: Platinum Jubilee lectures, 82nd Indian science congress, Part III: Engineering and Earth Sciences, pp 13–20
106. Gupta HK, Malomo S (1995) The Malawi earthquake of March 10, 1989: report of field survey. Seismol Res Lett 66(1):20–27
107. Gupta HK, Khanal KN, Upadhyay SK, Sarkar D, Rastogi BK, Duda SJ (1995) Verification of magnitudes of Himalayan region earthquakes of 1903–1985 from Goettingen observatory. Tectonophysics 244:267–284
108. Gupta HK (1995) 'Introduction' for the Uttarkashi Earthquake Volume, Memoir 30, Geol Soc India, pp v xiii
109. Gupta HK (1995) 'Latur Earthquake'. Vigyan Garima Sindhu, vol 19, Vaigyanik Tatha Takniki Shabdhavali Aayog, Min. Human Res. Dev., Govt of India Publication, pp 29–32
110. Gupta HK, SK Verma (1995). 'India's contributions to geophysical investigations in Antarctica', India & Antarctica during the Precambrian. Mem Geol Soc India 34:293–310
111. Gupta HK, Sarma SVS, Harinarayana T, Virupakshi G (1996) Fluids below the hypocentral region of Latur earthquake, India: geophysical indicators. Geophys Res Lett 23(13):1569–1572
112. Gupta HK, Dwivedy KK (1996) Drilling at Latur earthquake region exposes a Peninsular Gneiss basement', short communication. Geol Soc India 47:129–131
113. Satyabala SP, Gupta HK (1996) Is the Quiescence of major earthquakes ($M \geq 7.5$) since 1952 in the Himalaya and Northeast India real? Bull of Seis Soc Am 86(6):1983–1986

114. Sato K, Bhatia SC, Gupta HK (1996) Three dimensional numerical modeling of deformation and stress in the Himalaya and Tibetan Plateau with a simple geometry. J Phys Earth 44:227–254
115. Gupta HK (1997) Meeting focuses on catastrophic Asian earthquakes. Asian Seismol Comm Rep EOS 78(14):147
116. Gupta HK, Chadha RK, Rao MN, Narayana BL, Mandal P, Ravi Kumar M, Kumar N (1997) The Jabalpur Earthquake of May 22, 1997. J Geol Soc India 50(7):85–91
117. Gupta HK et al (1997) Reply to the comments by Dr. S. K. Acharya, Director General, GSI on the paper "The Jabalpur Earthquake of May 22, 1997". J Geol Soc India, 50(3):375–376
118. Gupta HK, Rastogi BK, Chadha RK, Mandal P, Sarma CSP (1997) Enhanced reservoir induced earthquakes in Koyna region, India, during 1993–95. J Seismolog 1(1):47–53
119. Duda SJ, Gupta HK (1997) Spectral magnitudes and seismograms measure radiation from seismic sources. Eos 78(39):417–423
120. Gupta HK, Bhatia SC (1997) Global Seismic Hazard Assessment Programme (GSHAP) and its relevance to India. In: Proceeding of workshop on earthquake disaster preparedness, 13–14 October, Roorkee
121. Chadha RK, Gupta HK, Kumpel HJ, Mandal P, Nageswara Rao A, Narendra Kumar I, Radhakrishna BK, Rastogi IP, Raju CSP, Sarma CS, Satyanarayana HVS (1997) Delineation of active faults, nucleation process and pore pressure measurements at Koyna (India). Pure Appl Geophys 150:551–562
122. Gupta HK (1997) Major and Great Earthquakes in the Himalayan Region. In: Proceedings of the workshop sponsored by the Department of Science & Technology, Govt of India, New Delhi and North Eastern Council, Govt of India, Shillong, The Great Shillong Earthquake 1897 A Centennial Retrospective, AEG Pub, no 38, pp 1–4
123. Gupta HK, Rastogi BK, Indra Mohan CVRK, Rao SVSS, Rao RUM (1998) An investigation into the Latur Earthquake of September 29, 1993 in Southern India. Tectonophysics 287:299–318
124. Gupta HK, Johnston AC (1998) NOTE on "Chapman Conference on Stable Continental Region (SCR) Earthquakes". J Geol Soc India. 52(1):115–117
125. Gupta HK, Johnston AC (1998) Stable continental regions are more vulnerable to earthquakes than once thought' (Chapman Conference Report). EOS Trans AGU 79(27):319–321
126. Kumpel HJ, Gupta HK, Radhakrishna I, Chadha RK, Grecksch G (1998) "Indo German research collaboration on in situ pore pressure studies in the Koyna Region, India", Extended Abstract, Proc. of Workshop: 'Geodynamical hazards associated with large dams' conducted by European Centre for geodynamics and seismology. Conseil De'L Europe 16:153–156
127. Gupta HK (1998) "Latur Bhookamp Par Ek Drishtipaath" Special Article, 'Antariksh 2000' Inter Centres Hindi Technical Seminar (23–24 Dec 1998) publication, ISTRAC, Dept of Space, GOI pp 1–6
128. Krishna VG, Rao CVRK, Gupta HK, Sarkar D, Baumbach M (1999) Crustal seismic velocity structure in the epicentral region of the Latur earthquake (September 29, 1993), southern India: inferences from modelling of the aftershock seismograms. Tectonophysics 304:241–255
129. Bhatia SC, Ravi Kumar M, Gupta HK (1999) A probabilistic seismic hazard map of India and adjoining regions. Ann Geofis 42(6):1153–1164
130. Zhang P, Yang ZX, Gupta HK, Bhatia SC, Shedlock KM (1999) Global seismic hazard assessment program (GSHAP) in continental Asia. Ann Geofis 42(6):1167–1190
131. Sukhija BS, Rao MN, Reddy DV, Nagabhushanam P, Hussain S, Chadha RK, Gupta HK (1999) Paleoliquefaction evidence and periodicity of large prehistoric earthquakes in Shillong Plateau, India. Earth Planet Sci Lett 167:269–282
132. Gupta HK, Bhattacharya SN, Ravi Kumar M, Sarkar D (1999) Spectral characteristics of the 11 May 1998 Pokhran and 28 May 1998 Chaghai nuclear explosions. Curr Sci 76(8):1117–1121
133. Gupta HK, Rao RUM, Srinivasan R, Rao GV, Reddy GK, Dwivedi KK, Banerjee DC, Mohanty R, Satyasaradhi YR (1999) Anatomy of surface rupture zones of two stable continental region earthquakes, 1967 Koyna and 1993, Latur, India. Geophys Res Lett 26(13):1985–1988

134. Singh SK, Ordaz M, Dattatrayam RS, Gupta HK (1999) A spectral analysis of the May 21, 1997 Jabalpur, India earthquake (Mw=5.8) and estimation of ground motion from future earthquakes in the Indian shield region. Bull Seism Soc Am 89:1620–1630
135. Gupta HK, Radhakrishna I, Chadha RK, Kumpel HJ, Grecksch G (2000) Pore pressure studies initiated in area of reservoir-induced earthquakes in India. EOS Trans 81(14):145–151
136. Gupta HK (2000) The Deccan: site of World's most prominent stable continental region earthquakes. In: Proceeding of INSA seminar on Deccan Heritage, Spl. Publication, Universities Press, pp 109–134
137. Gupta HK (2000) Major and great earthquakes of the Himalayan Region: an overview. In: Earthquake hazard and seismic risk reduction. Kluwer Academic Publishers, pp 79–85
138. Gupta HK (2000) Earthquake hazard in developing countries and GSHAP (Global seismic hazard assessment programme). In: Proceeding of second international workshop on 'Earthquakes and Megacities' held during 1–3 Dec, 1999 at Makati City, Philippines, pp 1–8
139. Gupta HK, Purnachandra Rao N, Rastogi BK, Sarkar D (2001) The deadliest intraplate earthquake. Science 16th Mar 2001 291:2101–2102
140. Rastogi BK, Gupta HK, Mandal P, Satyanarayana HVS, Kousalya M, RaghavanR, Jain R, Sarma ANS, Kumar N, Satyamurty C (2001) The deadliest stable continental region earthquake occurred near Bhuj on 28 January 2001. J Seism, Oct 2001 5(4):609 (Kluwer Academic Publishers)
141. Gupta HK (2001) Medium term forecast of the 1988 north-east India earthquake. Elsevier, Tectonophysics 338(3–4):281–286
142. Gupta HK (2001) Short-term earthquake forecasting maybe feasible at Koyna, India. Elsevier, Tectonophysics 338(3–4):353–357
143. Gupta HK, Harinarayana T, Kousalya M, Mishra DC, Mohan I, Purnachandra Rao N, Raju PS, Rastogi BK, Reddy PR, Sarkar D (2001) Notes on "Bhuj Earthquake of 26th January, 2001". J Geol Soc India 57
144. Gupta HK, Mandal P, Rastogi BK (2002) How long will triggered earthquakes at Koyna, India continue? Curr Sci 82(2):202–210
145. Gupta HK (2002) A review of recent studies of triggered earthquakes by artificial water reservoirs with special emphasis on earthquakes in Koyna, India. Earth-Sci Rev 58(3–4):279–310
146. Gupta HK (2002) Oldest Neolithic settlements discovered in Gulf of Cambay. J Geol Soc India 59:277–278
147. Gupta HK, Srinivasan R, Rao RU, Rao GV, Reddy GK, Roy S, Jafri SH, Dayal AM, Zachariah J, Parthasarathy G, Rao P, Gowd GVS, Rao TN, Srirama SV, Dwivedy KK, Banerjee DC, Mohanty R, Satyasaradhi YR, Ksyyi VJ, Prasad AR, Ramanujam CGK, Reddy PR, Manoj S (2003) Borehole investigations in the surface rupture zone of the 1993 Latur SCR Earthquake, Maharashtra, India overview of results. J Geol Soc India 54:1–22
148. Gupta HK (2003) IOC Bruun Memorial Lectures, 2003. "Gas-Hydrates: A Potential Source of Energy from the Oceans", IOC Technical Series 65. UNESCO 2004:1–16
149. Gupta HK (2003) Major and great earthquakes in the Himalayan Region: an overview. 'Disaster Management', Universities Press (India) Pvt. Ltd., pp 100–109
150. Gupta HK (2003) "NATURE OF EARTHQUAKES", in natural and human induced hazards, Encyclopedia of Life Support Systems (EOLSS), developed under the auspices of UNESCO. Eolss Publishers, Oxford, UK, p 42
151. Gupta HK (2005) A note on the 26 December 2004 Tsunami in the Indian Ocean. J Geol Soc India 65(2):247
152. Gupta HK (2005) Mega-Tsunami of 26th December, 2004: Indian initiative for early warning system and mitigation of oceanogenic hazards. Episodes, March 2005
153. Gupta HK (2005) Artificial water reservoir-triggered earthquakes with special emphasis at Koyna. Curr Sci 88(10):1628–1631
154. Singh SK, Dattatrayam RS, Suresh G, Iglesias A, Bansal BK, Xyoli Perez M, Ordaz HK, Gupta PR, Baidya JL, Gautam GK, Singh RK (2005) The Great Sumatra—Andaman Earthquake of 2004: Regional Broadband Seismograms from India. Seismol Res Lett 76(6):684–692

155. Gupta HK, Mandal P, Satyanarayana HVS, Shashidhar D, Sairam B, Shekar M, Singh A, Uma Devi E, Kousalya M, Purnachandra Rao N, Dimri VP (2005) An earthquake of M~5 may occur at Koyna. Curr Sci 89(5):747–748
156. Gupta HK, Mandal P, Satyanarayana HVS, Shashidhar D, Sairam B, Shekar M, Singh A, Uma Devi E, Kousalya M, Purnachandra Rao N, Dimri VP (2005) Earthquake at Koyna. Curr Sci 89(11):1785–1787 10 Dec 2005
157. Singh SK, Oritz M, Gupta HK, Ramadass DGA (2006) Slow slip below Port Blair, Andaman, during the great Sumatra-Andaman earthquake of 26 December 2004. Geophys Res Lett 33(L03313):1–4
158. Gupta HK, Shashidhar D, Periera M, Mandal P, Dimri VP (2006) Prediction of an M~4 Earthquake in the Koyna Region Comes True! J Geol Soc India 68:149–150
159. Gupta HK, Shashidhar D, Pereira M, Mandal P, Purnachandra Rao N, Kousalya M, Satyanarayana HVS, Dimri VP (2007) Earthquake forecast appears feasible at Koyna, India. Curr Sci 93(6):843–848
160. Gupta HK, Shashidhar D, Pereira M, Purnachandra Rao N (2007) A new zone of seismic activity at Koyna, India. J Geol Soc India 69:1136–1137
161. Gupta HK, Shashidhar D, Mallika K, Purnachandra Rao N, Srinagesh D, Satyanarayana HVS, Saha S, Naik RTB (2007) Short term forecast for an M~4 earthquake at Koyna, India. J Geol Soc India 70, Dec 2007, p 1091
162. Sain K, Gupta HK (2008) Gas-hydrates-future potential source of energy in India. Glimpses Geosci Res India, pp 244–250
163. Sain K, Gupta HK (2008) Gas hydrates: Indian scenario. J Geol Soc India 72:299–311
164. Gupta HK, Purnachandra Rao N, Shashidhar D, Mallika K (2008) The disastrous M 7.9 Sichuan Earthquake of 12 May 2008. J Geol Soc India 72:325–330
165. Gupta HK (2008) "Artificial water reservoir triggered earthquakes, with special emphasis on Koyna Earthquakes, India", Memoir 66. J Geol Soc India,1–28p
166. Gupta HK (2008) Two significant developments in seismology in India: study of artificial water reservoir triggered earthquakes, and Tsunami hazard mitigation, Memoir 68. Golden Jubilee Volume of Geol Soc India, 15–30
167. Gupta HK (2008) India's initiative in mitigating Tsunami and storm surge hazard. J Earthquake Tsunami 2(4):287–295
168. Gupta HK, Gahalaut VK (2009) Is Northern Bay of Bengal Tsunamigenic. Bull Seismol Soc Am 99(6):3496–3501
169. Shashidhar D, Kumar N, Mallika K, Gupta HK (2010) Characteristics of seismicity patterns prior to the M~5 earthquakes in the Koyna Region, Western India-application of the RTL algorithm. Episodes, J Int Geosci 33:2 (June 2010):83–90
170. Gupta HK (2010) Co-operation plan on hazards and disasters risk reduction in Asia and the Pacific. In: Beer T (ed) Geophysical hazards, minimizing risk, maximizing awareness; International year of planet earth, https://doi.org/10.1007/978-90-481-3236-2_3. © Springer Science + Business Media B.V.2010
171. Gupta HK (2010) India's initiative in mitigating Tsunami and storm surge hazard. Science and science policy. In: Proceedings of the concluding meeting of the Platinum Jubilee Celebrations, Indian National Science Academy, 7–10 Dec 2009, Kolkata, p 107
172. Shashidhar D, Purnachandra Rao N, Gupta HK (2011) Waveform inversion of broad-band data of local earthquakes in the Koyna-Warna region, western India. Geophys J Int. 1–13. https://doi.org/10.1111/j.1365-246X.2011.04935.x
173. Gupta HK, Shashidhar D, Mallika K, Purnachandra Rao N, Srinagesh D, Satyanarayana HVS, Saha S, Naik RTB (2011) Short term earthquake forecasts at Koyna, India. J Geol Soc India 77:5–11 (Jan 2011)
174. Gupta HK (2011) Artificial water reservoir triggered earthquakes, Encyclopedia of solid earth geophysics. Springer+ Business media, 1, pp 15–24
175. Gupta HK (2011) Planning a deep drilling project in the Koyna Region of India, workshop on deep scientific drilling to study reservoir triggered earthquakes, Koyna, India, 21–25 Mar 2011. Eos 92:283

176. Gupta HK, Nayak S (2011) Deep scientific drilling to study reservoir-triggered earthquakes in Koyna, Western India. Sci Drill 12:53–54
177. Sain K, Gupta HK (2012) Gas hydrates in India: potential and development. Gondwana Res 22:645–657
178. Mallika K, Gupta HK, Shashidhar D, Purnachandra Rao N, Yadav A, Sunil Rohilla, Satyanarayana HVS, Srinagesh D (2013) Temporal variation of b value associated with M ~4 earthquakes in the reservoir-triggered seismic environment of the Koyna-Warna region, Western India. J Seismol 17:189–195
179. Roy S, Rao NP, Akkiraju VV, Goswami D, Sen M, CSIR-National Geophysical Research Institute, Hyderabad; Gupta HK, National Disaster Management Authority, Govt. of India, New Delhi and Bansal BK, Nayak S, Ministry of Earth Sciences, Govt. of India, New Delhi (2013) Granitic Basement below Deccan Traps Unearthed by Drilling in the Koyna Seismic Zone, Western India. J Geol Soc India 81:281–290 pp
180. Srinivasa Kumar T, Nayak S, Gupta HK (2013) India's Tsunami early warning system, extreme natural hazards, disaster risks and societal implications. Cambridge University, pp 322–330
181. Gupta HK (2013) Triggered earthquakes, Encyclopedia of natural hazards, encyclopedia of earth science series. Springer, pp 1031–1036
182. Gupta HK, Sain K (2013) Gas-Hydrates, Encyclopedia of natural hazards, Encyclopedia of earth science series. Springer, pp 377–378
183. Gupta HK, Gahalaut VK (2014) Seismotectonics and large earthquake generation in the Himalayan region. Gondwana Res 25(1):204–213
184. Shashidhar D, Mallika K, Rao NP, Gupta HK, Satyanarayana HVS (2014) Detection of quarry blasts in the Koyna-Warna region, western India. OJER 2014(3):162–169
185. Gupta HK, Nayak S, Ellsworth W, Rao YJB, Rajan S, Bansal BK, Purnachandra Rao N, Roy S, Arora K, Mohan R, Tiwari VM, Satyanarayana HVS, Patro PK, Shashidhar D, Mallika K (2014) Probing reservoir-triggered earthquakes in Koyna, India, through scientific deep drilling. Sci Drill 18:5–9
186. Gupta HK, Purnachandra Rao N, Roy S, Arora K, Tiwari VM, Patro BPK, Satyanarayana HVS, Shashidhar D, Mallika K, Akkiraju VV, Goswami D, Vyas D, Ravi G, Srinivas KNSSS, Srihari M, Mishra S, Dubey CP, Raju ChV, Ujjal Borah K, Reddy C, Babu N, Rohilla S, Dhar U, Mrinal Sen, Rao BYJ (2014) Investigations related to scientific deep drilling to study reservoir-triggered earthquakes at Koyna, India. Int J Earth Sci 104:1511–1522
187. Gupta HK, Gahalaut VK (2015) Can an Earthquake of Mw~9 occur in the Himalayan Region? Geol Soc London, Special Publications 412:43–53
188. Cutter SL, Zadeh AI, Alcantara-Ayala, Altan O, Baker DN, Briceno S, Gupta HK, Holloway A, Johnston D, McBean GA, Ogawa Y, Paton D, Porio E, Silbereisen RK, Takeuchi K, Valsecchi GB, Vogel C, Wu G (2015) Pool knowledge to stem losses from disasters. Nature 522:277–279
189. Shashidhar D, Satyanarayana HVS, Mahato CR, Mallika K, Purnachandra Rao N, Gupta HK (2016) Borehole seismic network at Koyna, India. Seismol Res Lett 87(3):661–667
190. Gupta HK, Kusumita Arora, Rao PN, Sukanta Roy, Tiwari VM, Patro BPK, Satyanarayana HVS, Shashidhar D, Mahato CR, Srinivas KNSSS, Srihari M, Satyavani N, Srinu Y, Gopinadh D, Raza H, Jana M, Akkiraju VV, Goswami D, Digant Vyas, Dubey CP, Raju DCV, Borah U, Kashi Raju, Reddy KC, Narendra Babu, Bansal BK, Nayak S (2016) Investigations of continued reservoir triggered seismicity at Koyna, India. Geol Soc London, Special Publications 445:151–188
191. Gupta HK, Shashidhar D, Mahato CR, Satyanarayana HVS, Mallika K, Purnachandra Rao N, Maity BS, Navitha K (2017) Location of the pilot borehole for investigations of reservoir triggered seismicity at Koyna, India. Gondwana Res, Elsevier 42:133–139
192. Goswami D, Akkiraju VV, Misra S, Roy S, Singh SK, Amalendu Sinha, Gupta HK, Bansal BK, Nayak S (2017) Rock strength measurements on Archaean basement granitoids recovered from scientific drilling in the active Koyna seismogenic zone, western India. Elsevier, Tectonophysics, pp 182–192

193. Misra S, Roy S, Bartakke V, Gaurav A, Gupta HK (2017) Fissures and fractures in the Koyna seismogenic zone, Western India. Geol Soc India 90(2):131–137
194. Gupta HK (2017) Koyna, India, an ideal site for near field Earthquake observations. Geol Soc India 90(6):645–652
195. Gupta HK (2018) Review: Reservoir Triggered Seismicity (RTS) at Koyna, India over the past 50 years. Bull Seismol Soc Am 108(5B):2907–2918
196. Gupta HK (2020) Recent earthquakes in Delhi and developing an earthquake resilient society. J Geol Soc India 96:1–4
197. Elliott JR, de Michele M, Gupta HK (2020) Earth observation for crustal tectonics and earthquake hazards. Surv Geophys 2(41):1355–1389. https://doi.org/10.1007/s10712-020-09608
198. Hebert H, Occhipinti G, Schindele F, Gailler A, Pinel-Puyssegur B, Gupta HK, Rolland L, Lognonne P, Lavigne F, Meilianga E, Chapkanski S, Crespon F, Paris A, Heinrich P, Monnier A, Jamelot A, Reymond D (2020) Contributions of Space Missions to Better Tsunami Science: Observations, Models and Warning. Surv Geophys 2(41)1535–1581 https://doi.org/10.1007/s10712-020-09616-2
199. Gupta HK, Sabnis KA, Duarah R, Saxena RS, Baruah S (2020) Himalayan earthquakes and developing an earthquake resilient society. J Geol Soc India 96:433–446
200. Goff J, Gupta HK, Glade T (2020) Editorial at the occasion of the 100th volume of natural hazards. Nat Hazards, Springer 100(1):1–2
201. Rekapalli R, Gupta HK (2021) Characterization of injection-induced seismicity at north central Oklahoma, USA. Short communication. J Seismolog 25:327–337
202. Gupta HK (2021) Understanding anthropogenic earthquakes. Curr Sci 120(9):1415–1416
203. Gupta HK (2021) Koyna, India, an ideal site for near field earthquake observations. J Geol Soc India 97(10):1144–1151
204. Gupta HK (2021) Studies of artificial water reservoir triggered earthquakes at Koyna, India: A summary. J Geol Soc India 97(12):1556–1564
205. Gupta HK, Sabnis KA (2021) Developing an earthquake resilient society in the vicinity of Himalaya. J Geol Soc India 97(12):1593–1602
206. Gupta HK (2022) Artificial water reservoir-triggered seismicity (RTS): most prominent anthropogenic seismicity. Surv Geophys,43:619–659 https://doi.org/10.1007/s10712-021-09675-z
207. Gupta HK (2022) Three different types of earthquake sequences in the Deccan Volcanic Province of India. Geosyst Geoenviron 1(4):1–5
208. Gupta HK (2023) If a magnitude ~8 Earthquake Occurs in India Today…. Journal of Geological Society of India 99:299–302
209. Rekapalli R, Gupta HK (2023). "How long the Mw $\geq$ 5 aftershocks of the 6 February 2023 Mw 7.8 Turkiye earthquake shall continue? Natural Hazards, 117(3): 3399–3402 https://doi.org/10.1007/s11069-023-05953-8.
210. Gupta HK (2023) Himalayan Seismic Belt, Seismic Gaps and Related Issues. Journal of Geological Society of India 99:1187–1190
211. Gupta HK, Rekapalli R (2023). "Duration of Mw $\geq$ 5 aftershocks of the 6 February 2023 Mw 7.8 Turkiye earthquake. Natural Hazards, 120(4)-4031-4033 https://doi.org/10.1007/s11069-023-06325-y.

Books Authored

212. Gupta HK, Rastogi BK (1976). 'Dams and Earthquakes', Elsevier Scientific Publishing Company, Amsterdam, 229 p. (Translated into Russian in 1979 and Chinese in 1980).
213. Gupta HK (1980) Geothermal Resources: An Energy Alternative. Elsevier Scientific Publishing Company, Amsterdam, 227p

214. Gupta HK (1992) Reservoir Induced Earthquakes. Elsevier Scientific Publishing Company, Amsterdam, 364p
215. Gupta HK, Roy S (2006) Geothermal Energy: An Alternative Resource for the 21st Century. Elsevier Scientific Publishing Company, Amsterdam, 279p
216. Gupta HK, Gahalaut VK (2013). Three Great Tsunamis: Lisbon (1755), Sumatra-Andaman (2004) and Japan (2011), Springer Briefs in Earth Sciences, Springer, ISBN: 978–94–007–6575–7, IX, 89p.
217. Gupta, HK(2021). "Building Dakshin Gangotri at Antarctica: A Miracle "Journal of Geological Society of India", ISBN: 978–9380998–37–4, 128p.

Encyclopedia of Solid Earth Geophysics

218. Gupta, HK (2011). Editor-in-Chief, Encyclopedia of Solid Earth Geophysics, Volume I and Volume II, published by Springer, Springer ISBN: 978–90–481–8701–0, p. 1–1539. (https://link.springer.com/referencework/https://doi.org/10.1007/978-90-481-8702-7).
219. Gupta, HK (2021). Editor-in-Chief, Encyclopedia of Solid Earth Geophysics, 2nd Edition, Volume I and Volume II, published by Springer, Springer ISBN: ISBN 978–3–030–58631–7, p. 1–1950. (https://link.springer.com/referencework/ https://doi.org/10.1007/978-3-030-586 31-7).

Volumes Edited

220. Gupta HK (1970) Seismology in India. Special Publication of the Indian Geophysical Union, Dr.A.N.Tandon Commemoration Volume, 165 p
221. Gupta HK (1973). Proceedings 'International Seminar on Geodynamics of the Himalayan Region', Inter Union Commission on Geodynamics, Sci. Report Series, No.5, 215 p.
222. Gupta HK, Delany FM(1981). 'Zagros, Hindu Kush Himalaya Geodynamic Evolution', American Geophysical Union, Geodynamics Series, 3, 323 p.
223. Naqvi SM, Gupta HK, Balakrishna S (1984). 'Lithosphere, Structure, Dynamics and Evolution', Elsevier Scientific Publishing Co., Tectonophysics Special Issue, 105, 419 p.
224. Gupta HK (1986). Scientific Report of Third Indian Expedition to Antarctica, Department of Ocean Development, Govt. of India, Technical Publication No.3, 266 p.
225. Gupta HK (1987). 'Deep Seated Processes in Collision Zones', Elsevier Scientific Publishing Company, Tectonophysics Special Issue, 134, No.1–3, 238 p.
226. Gupta HK(1989). 'Seismological Instrumentation and Data Analysis in Developing Countries', Physics of the Earth and Planetary Interior, Amsterdam, 58, 353 p.
227. Gupta HK (1992). 'Seismology in India an Overview' Current Science, Special Issue, Vol.62, No.1&2, 1992, 264 p.
228. Ebinger CJ, Gupta HK, Nyambok OI (1992). 'Seismology and Related Sciences in Africa', Elsevier Scientific Pub. Co., Tectonophysics, Special Issue, Vol. 209, No. 1–4, 337 p.
229. Gupta HK (1994). 'Latur Earthquake', Memoir 35, Geol. Soc. of India, 149 p.
230. Gupta HK (1995). 'Uttarkashi Earthquake' Memoir 30 Geol. Soc. of India, 233 p
231. Gupta HK, Chadha RK (1995) Induced Seismicity. PAGEOPH (Topical Volume) 145(1):217
232. Gupta HK, Parasher-Sen A, Balasubramanian D (2000). "Deccan Heritage", Proc. of INSA Seminar on Deccan Heritage, Spl. Publication, ISBN 81–7371–285–9, 253 p.
233. Gupta HK, Chadha RK, Srinagesh D (2001),"The Nature of Seismic Sources and the Prediction of Earthquakes", Elsevier Scientific Pub. Co., Tectonophysics, Special Issue, Vol.338, Issues 3–4, 360 p.

234. Gupta HK, Gupta GD (2003), "Earthquake Studies in Peninsular India - Since 1993", Memoir 54, Geological Society of India, ISBN: 81–85867–54–2, 254 p.
235. Gupta HK (2003). "Disaster Management", Universities Press (India) Pvt. Ltd., ISBN 978-81-7371-456-6,152 p.
236. Gupta HK (2004). "Oceanology", Universities Press (India) Pvt. Ltd., ISBN 978-81-7371-529-7, 222 p
237. Gupta HK (2005). "Great Tsunami 26 December, 2004 Sumatra Region", Geological Society of India, ISBN: 81-85867-70-4,144 p.
238. Gupta HK, Chadha RK (2006) "Stable Continental Region Earthquakes" Special Issue of the Journal of Indian Geophysical Union, 62 p.
239. Gupta HK, Fareeduddin (2008) "Recent Advances in Earth System Sciences", Golden Jubilee Volume of Geological Society of India, Memoir 66, 680 p.
240. Gupta HK, Purnachandra Rao N, Yeats R (2009) "The Devastating Muzaffarabad earthquake of 8 October 2005 in Western Himalaya", Journal of Seismology, Vol.13, No.3, July 2009, ISSN 1383–4649, 420 p.
241. Gupta HK (2017) "50 years of 10 December 1967 M 6.3 Koyna Earthquake", Journal of Geological Society of India, Vol.90, No. 6, Dec 2017, 637–812.
242. Gupta HK (2019) "1919-2019 Centennial Celebrations of the International Union of Geodesy and Geophysics(IUGG): Contributions from India." Proceedings of Indian National Science Academy 85(2):315–504
243. Tiwari VM, Gupta HK (2021) '60 Years of CSIR- National Geophysical Research Institute'. Journal of Geological Society of India 97(10):1113–1324
244. Gupta HK, Wu Z, Banerjee P, Li L (2021) 25 Years of Asian Seismological Commission. Journal of Geological Society of India 97(12):1475–1602
245. Kalachand Sain, Sukanta R, Gupta HK (2022). "Emerging Energy Resources in India". Journal of Geological Society of India, 222 p.
246. Science Plan on Hazards and Disasters", (Earthquakes, Floods and Landslides) of ICSU-ROAP JUNE 2008. (https://council.science/wp-content/uploads/2017/05/ICSU-ROAP-Science-Plan-Hazards-and-Disasters.pdf)
247. "Science Plan on Hazards and Disasters", (Special Vulnerability of Islands) of (ICSU-ROAP) JUNE 2008. (https://council.science/publications/science-plan-on-hazards-and-disasters-special-vulnerability-of-islands-icsu-regional-office-for-asia-the-pacific/)

Printed by Printforce, the Netherlands